TRANSACTIONS

of the

American Philosophical Society

Held at Philadelphia for Promoting Useful Knowledge

VOLUME 77, Part 4, 1987

Royal Funding of the Parisian Académie Royale des Sciences During the 1690s

ALICE STROUP

THE AMERICAN PHILOSOPHICAL SOCIETY

Independence Square, Philadelphia

1987

Library of Congress Catalog
Card Number 86-71785
International Standard Book Number 0-87169-774-2
US ISSN 0065-9746

To my parents
Gloria B. Tyler and Parker R. Tyler, Jr.

CONTENTS

ILLUSTRATIONS

Plates

Figures

TABLES

ACKNOWLEDGMENTS

When I first began ten years ago to investigate royal funding of the early Académie royale des sciences, it quickly became apparent that the data for the end of the seventeenth century were completely unknown and that without them meaningful analysis would be impossible. Thanks to a grant from the American Philosophical Society, I was able more than five years ago to begin searching for the missing information. At the time, I envisaged this research as contributing to a chapter in my larger study of the Academy, *A Company of Scientists.* Such was my optimism that I anticipated neither the elusiveness of the evidence, nor the difficulties in understanding the documents in which it was buried, nor the extent to which the subject merited an extended analysis. This book has been an unexpected, indeed accidental, by-product.

In the course of my research many individuals and institutions have given valuable help; I am grateful for their advice and support. If they scarcely recognize the early drafts in this final product, that is because their suggestions were taken to heart. Grants from the American Philosophical Society, the National Endowment for the Humanities, the American Council of Learned Societies, and the National Science Foundation enabled me to research the Academy's finances in Parisian archives during the summers of 1981, 1982, and 1983, and to begin writing this book. A grant from Bard College provided support during the writing of chapter 4.

Parts of chapters 3 and 6 were presented at the annual meeting of the American Historical Association in Los Angeles, December 1981, and to the Metropolitan Seminar for the History of Technology (New York); discussion in both forums clarified my argument. I am grateful to Professors Roger Hahn, Bert Hansen, Michael S. Mahoney, and R. S. Westfall, and to the referees of the National Science Foundation for their criticisms of early drafts; to Mme Aline Vallée and Professor Bernard Barbiche, for their valuable advice about the archival record; and to Professor Mahoney, who generously shared with me his own findings and opinions about the finances of the Academy. Special thanks are due to my colleague Professor John C. Fout for his encouragement at all stages of the work.

The staffs of various libraries and archives were helpful. I particularly wish to thank Jane Hryshko at the Bard College Library and Margareta Blumenthal at Åbo Akademis Bibliotek, who ordered essential materials from other libraries; M. Pierre Berthon and Mme Claudine Pouret at the Archives of the Académie des Sciences; and the presidents of the Salle Clisson at the Archives Nationales and of the Département des Manuscrits at the Bibliothèque Nationale. I am grateful to the Permanent Secretaries of the Académie des Sciences of the Institut de France for permission to quote from manuscripts in the Archives of the Académie des Sciences, and to the Bibliothèque Nationale for permission to reproduce the portraits of Pontchartrain and Bignon.

It is a pleasure to thank the editors at the American Philosophical Society for their patience and sound advice, Kristina Mickelson for helping with the typing, Willie G. Pannell, Jr., who checked the tables in their original form, and the Bureau d'Accueil des Professeurs d'Universités Etrangères of the Université de Paris which provided facilities for revisions of the tables in a later form.

This is not the last word on the subject, but I hope it will be useful to future students of learned institutions or royal expenditure during the reign of Louis XIV. I have tried in the references and appendices to indicate where the archival record will reward further investigation, and I apologize for any errors of fact or judgment, for which I bear sole responsibility.

ABBREVIATIONS USED IN THE NOTES

Complete citations for all references in the footnotes can be found in the bibliography.

AdS	Paris, Académie des Sciences, Archives.
AdS, Reg.	Paris, Académie des Sciences, Archives, Registre des procès-verbaux des séances.
AN	Paris, Archives Nationales.
BN	Paris, Bibliothèque Nationale.
CdB	*Les comptes des bâtiments du roi sous le règne de Louis XIV*, ed. Guiffrey.
DBF	*Dictionnaire de biographie française*, ed. Balteau.
DSB	*Dictionary of Scientific Biography*, ed. Gillispie.
Histoire	Fontenelle, *Histoire de l'Académie royale des sciences (1666–99)*.
Histoire . . . (date)	Académie royale des sciences, *Histoire . . .*, avec les mémoires . . . (1699–1790); see the section "Histoire."
Historia	Du Hamel, *Regiae Academiae Scientiarum Historia*.
IB	Institut de France, Académie des Sciences, *Index biographique des membres et correspondants de l'Académie des sciences*.
Mémoires	Vols. 3–11 of Académie royale des sciences, *Histoire et mémoires . . . depuis 1666 jusqu'à 1699*.
Mémoires . . . (date)	Académie royale des sciences, *Histoire . . .*, avec les mémoires . . . (1699–1790); see the section "Mémoires."
NBU	Hoefer, ed., *Nouvelle biographie universelle*.

Other abbreviations are explained in the Key to the Tables and in appendix A.

I. INTRODUCTION

The scientific revolution of the seventeenth century engendered diverse and prolific offspring, among which were the scientific societies. Some of these institutions were the spontaneous embodiment of scientists' desires to share ideas in formal meetings. The French Académie royale des sciences, however, offers distinct and suggestive contrasts to the general pattern. Founded in 1666 by Jean-Baptiste Colbert, Louis XIV's minister of finance, the Academy was the beneficiary of the most generous patronage of science known during the seventeenth century. It was an official, governmental expression of support for science rather than the independent, scholarly coterie characteristic of other contemporary scientific societies. It was a creature not only of the scientific revolution but also of the French state.

The accomplishments of the Academy of Sciences can be appreciated directly in its discoveries and publications. Taken together these present a picture of scientific thought during the period and of the interchange of ideas both among French savants and internationally. Another way of illuminating this history of ideas, while also explicating the nature and effects of patronage, is to study the Academy's funding, which not only supported scientific research but also gave the French government a means of controlling the institution. The finances of the early Academy clarify the research and organization of the fledgling institution and the policies of its three ministerial protectors during the seventeenth century—Colbert, Louvois, and Pontchartrain. The amount and nature of spending varied from decade to decade, depending on the interests of academicians, the state of the treasury, and the enthusiasm of each minister. Yet there were common themes as well: for Colbert, founding the Academy entailed subsidizing it, and his successors inherited his Academy, his policies, and to some extent his theory of patronage.

Analysis of the Academy's finances can also correct distortions of perspective. Ever since Fontenelle wrote his history of the early Academy, Colbert has been given credit for encouraging the sciences for their own sake, while Louvois was thought to be interested chiefly in practical applications, and Pontchartrain is remembered for trying to revive a demoralized institution. However, a more balanced picture, taking account of the finances, would also show Colbert as favoring technological research and projects directly benefiting the king, Louvois as intending to publish works in natural history, and Pontchartrain as expanding the Academy and regularizing its affairs.[1]

---

[1] Stroup, *Company*, chaps. 4–5.

This work will focus on Louis Phélypeaux de Pontchartrain, Louis's minister of finances during the 1690s and third protector of the Academy, the minister whose attitudes toward scholarship and patronage most resembled those of Colbert.[2] Colbert's and Louvois's control of the Academy are already discussed in my more comprehensive study of the early Academy, *A Company of Scientists*. But it is important to emphasize Pontchartrain's regime,[3] since both as minister and as protector of the Academy he has languished in the shadow of his predecessors. His administration has been neglected and misrepresented, and it has even been thought that Pontchartrain did not fund the Academy of Sciences at all. With the discovery of the crown's subsidies of the Academy during the 1690s, it is now possible to clarify Pontchartrain's patronage at a crucial time in the Academy's history, the decade immediately before the reorganization of 1699.

Discovering and analyzing the financial record of the Academy are more difficult than commonly realized. The problems are not so great for the early period, because when Colbert established the institution he charged its expenses to the king's buildings account, which he controlled, and these data have been published. When Colbert died, François Michel Le Tellier, marquis de Louvois assumed responsibility for the Academy and continued to charge its expenses to the buildings account until 1691. In 1690 and 1691 Louis reorganized the royal bureaucracy; as a result, when Pontchartrain became protector of the Academy in 1691, he charged practically no expenses on its behalf to the buildings account. Thus when Jules Guiffrey edited that account at the end of the nineteenth century, he concluded from the near disappearance of expenditure on the Academy after 1690 that support for the institution had been suspended or transferred to another account.[4]

Although the conjecture that the crown no longer funded the Academy is mistaken, it is plausible. Royal monetary support of scientific societies was far from the rule in the seventeenth century, and in light of the extreme financial difficulties of the 1690s, it would have been reasonable for the French crown to halt such an expense, at least temporarily. During the reign of Louis XIV, the kingdom suffered from a general recession and, after 1688, from the effects of a progressive devaluation of its currency.[5] The wars of the 1690s, combined with the terrible harvests of 1693-94 and exceptional expedients for increasing royal income, further weakened the economy. This was a decade dominated by a war economy, monetary in-

---

[2] Pontchartrain controlled the academies as part of the portfolio which fell to him as *secrétaire d'état* for the Maison du roi, a post he retained after Chamillart replaced him as *contrôleur général des finances: Histoire . . . 1699*, 2; Leibniz, *Oeuvres*, 1: 289. He genuinely respected scholarship: Boislisle, *Correspondance des contrôleurs généraux*, 1: xiii–xiv.

[3] Patrice Berger has begun the task of reevaluating Pontchartrain: see his "French Administration in the Famine of 1693," "Pontchartrain and the Grain Trade during the Famine of 1693," and "Rural Charity in Late Seventeenth Century France."

[4] *CdB*, 3: iv. Wolf, *Observatoire*, 155, 209. On the transition from Colbert to Claude le Peletier, see BN MS. fr. 7750: 72r–73r.

[5] Braudel and Labrousse, *Histoire*, 2: 329–65; Goubert, *Louis XIV*, 209, 294.

flation, and an abrupt rise in prices, so that most sectors of French society suffered.[6] The contracting economy affected artistic and intellectual circles. For example, during the 1690s the crown suppressed the royal tapestry-making companies, and throughout Louis's reign printers were reluctant to print erudite works or books in large formats, preferring to turn out small books for the popular trade. As Henri-Jean Martin has shown, these are signs of not only an economic but also an intellectual recession, and both were exacerbated by the crisis of the 1690s.[7]

At the same time, the subtle and controversial balance among social orders was shifting. At the end of the seventeenth century, as Louis XIV expanded his military operations and became more dependent on the nobility of the sword, the middle class had less opportunity for social advancement.[8] Since savants, natural philosophers included, came principally from the middle class and their work appealed more to the nobility of the robe than of the sword,[9] the development of French natural philosophy was further endangered. For the king to turn his back on the Academy of Sciences at a time of military and economic crisis would thus have been consistent with his larger social policy at the end of the century.

Parisian gossip surviving from the 1690s also supports the view that the Academy was no longer funded, for it speaks of the king's failure to pay academicians. Thus the marquis de L'Hospital, admitted to the Academy in June 1693 as an honorary member, explained that Pontchartrain had refused to appoint new academicians because the wars made payment even of existing members impossible.[10] By recruiting L'Hospital without a pension, Jean-Paul Bignon, president of the Academy, effectively modified Pontchartrain's rule against new appointments and put into practice his own policy that titled or wealthy savants should not be pensioned.[11] Once academician Pierre Varignon secured for himself the pension of a deceased colleague, he considered vacating his original chair in mathematics, which brought him less income, in favor of Jean Bernoulli; he decided, however, to wait because academicians were not receiving their pensions anyway.[12]

---

[6] Meuvret, "Les mouvements des prix"; Hauser, *Recherches et documents,* 24–34; Braudel and Labrousse, *Histoire,* 2: 359–61; Schaeper, *Economy,* 12–13; Goubert, *Louis XIV,* 204–206, 209, 212, 219. Lister, *Journey,* 80, 138, 145, 162–63; Malet, *Comptes rendus,* 105–107; Forbonnais, *Recherches,* 2: 59–60; and the memoir written by the abbé Pegere in BN MS. fr. 7750: 1–41.

[7] Martin, *Livre,* 914–15, 917–20, 959–60, 964–65. On the effects of censorship, see Neveu, "La vie érudite," 458–59, 461.

[8] Grassby, "Social Status."

[9] Martin, *Livre,* 652–53, 964. Brunot, *Histoire de la langue française,* 4,1: 406–17.

[10] Bernoulli, *Briefwechsel,* 1: 173, 190. I am grateful to Professor Michael S. Mahoney for drawing my attention to the Bernoulli correspondence.

[11] An honorary member received no pension. Bignon believed that pensions should serve as encouragement and payment for work, and that academicians with a name or fortune should not receive pensions: BN MS. Clairambault 566: 187–89.

[12] Bernoulli, *Briefwechsel,* 1: 190, 191; note the use of the word "chaire" to describe a position at the Academy. For Varignon's pension, see table 1; it began only when Cusset's stopped. Boileau and Racine corresponded about the difficulty of getting paid in June 1693: Racine, *Oeuvres,* 7: 92–98. Boileau addressed Bignon on the question of Racine's money, and Bignon promised to convey his arguments to Pontchartrain.

A rumor even circulated that when academicians met in December 1695 to honor the king and to ask Pontchartrain for their salaries, they were rebuffed when he inquired what work they were doing.[13]

Against the theory that royal funding for the Academy ceased after 1690 stands the entire Colbertian legacy of patronage, which even Louvois perpetuated to some extent. Colbert had originally funded the Academy out of a desire to keep his monarch abreast with the English, and on the basis of an incorrect report that Fellows of the English Royal Society had "the support and the purse of the king,"[14] but he and Louis had other reasons for continuing the patronage. Colbert tended to think along interventionist lines in industry and the arts, and successful experiments with subsidies for individual savants encouraged him to expect that a more controlled undertaking on a larger scale would have even more satisfactory results.[15] His "anti-recessionary policy" had both economic and intellectual beneficiaries, and he supported the arts, sciences, and letters, just as he did trade and industry, by establishing and subsidizing enterprises and encouraging their productivity. In the intellectual sphere the crown enjoyed some success, for books by academicians swelled the number of titles in natural philosophy during the period from 1670 to 1699.[16] In many respects Colbert's policy toward the Academy resembled his treatment of the royal and state manufactories.

The Academy also served in the royal program of censorship, incentives, and propaganda. Establishing royal institutions offered a means of controlling the kinds of work done in various fields. The crown discouraged publication of Jansenist literature, polemical essays in defense of Colbert's predecessor Fouquet, libertine works, or satires of the king and royal family. Aspirants to royal patronage muted their criticism of the reign.[17] In the campaign to propagandize for the reign, the Academy of Sciences was one

---

[13] BN MS. fr. 13070: 185. This rumor was recorded by père Léonard de Sainte Catherine, a discalced Augustinian and librarian to the monastery at the place Notre Dame des Victoires during the 1690s. He nurtured close ties to members of the republic of letters, and his notes record conversations with librarians, booksellers, professors, and other monks. Two of Léonard's informants, Jean Boivin and Nicolas Clément, were associated with the Bibliothèque du roi, where the Academy met from 1666 to 1699. On Léonard, see Neveu, "La vie érudite," which, however, does not discuss this note about the Academy. The remainder of the note refers to the Academy's work on the map of France, which Léonard pointed out had already cost the king 20,000 livres.

[14] Chapelain wrote this to Huet on 3 July 1665: Chapelain, Lettres, 2: 406; Collas, Jean Chapelain, 383, n. 4. Chapelain had earlier written to Carrel de Sainte-Garde that the English king had established "un fond fixe" for "la subsistance et les frais" of the Royal Society, which would prevent its destruction. Chapelain went on, "On en a déjà vu d'heureux et de considérables effets": Chapelain, Lettres, 2: 348; Collas, Jean Chapelain, 386, n. 2.

[15] Colbert, Lettres, 5: 600–601, 603, 609. Middleton, Experimenters, 37. Huygens, Oeuvres, 4: 416–17. Martin, Livre, 667, 669–70; the task of praise was distasteful to many authors. Stroup, Company, chap. 4.

[16] Martin, Livre, 662–66, 669–70, 673, 856–83, 961–62. This was to some extent an illusory accomplishment, however, for many of the books were published by the Imprimerie royale, whose volumes were distributed more by presentation than by sale. Leibniz complained that he did not own them because bookshops did not carry them: Leibniz, Oeuvres, 1: 309.

[17] Martin, Livre, 670.

of many weapons to enhance the king's reputation. Royal academies were more susceptible to the royal will than were universities, guilds, or private societies.[18] Although this was especially true in the arts, the Academy of Sciences was a tool of royal policy favoring innovation in medicine and the sciences,[19] and its publications redounded to the *gloire* of the king.

Ultimately, the alliance of scientists and the state was based, like all patronage, on mutual benefit. Savants needed economic support not only for themselves but also for their research. Natural philosophical research was becoming more costly, requiring newly invented apparatus such as the thermometer, barometer, air pump, telescope, and microscope, which savants could not always furnish from their own resources. As academicians they enjoyed access to equipment and supplies for experiments and, at the individual level, many received pensions and lodging. The royal treasury subsidized their publications and funded research trips. Even persons outside the Academy benefited: instrument-makers gained a fresh market for their wares, and mathematicians became salaried assistants for special projects. Laborers and entrepreneurs in the building trade found employment on the construction of the Observatory. The material benefits of association with a funded scientific society were substantial.

Princes wanted to know the size of their domains and needed maps; they hoped that trade and military benefits would follow from the discovery of a means of determining longitude at sea. Louis quizzed members of the Academy about geography and cartography, reviewed their new map of the kingdom, and drew on their expertise in supplying and beautifying his royal palaces. Academicians reviewed military inventions, studied strategy, and mapped the very coasts of Europe along which so many naval engagements were played out in the 1680s and 1690s. When illness endangered the king's life, his ministers emphasized the Academy's responsibility for improving medicine. To scholars and patrons alike natural philosophical research promised practical applications, and the improvement of technology and medical care was an esteemed goal, shared by prince, minister, and savant alike. Thus, founding and financing the Academy formed sound policy for Louis, while for members the advantages of the institution were many.[20]

All of these elements entered into the rationale for state funding of the Academy, a lesson which could not have been lost on Pontchartrain. Pontchartrain was known specifically for his efforts to rejuvenate the Academy after Louvois's deleterious protectorship. He named his nephew,

---

[18] This sentence is a close paraphrase of Nicolas Pevsner's statement quoted by Yates, *French Academies*, 300.

[19] Royal policy favoring innovation in medicine, medical teaching, and scientific thought can be traced through Renaudot's Bureau d'adresse, the Jardin royal, and the *Journal des sçavans*: Brown, *Scientific Organizations*, 13–30, 195, 263; Handford, "Chemistry at the Jardin du roi," 19–20, 47–50; Partington, *History of Chemistry*, 2: 172–73, 269, 289; Brygoo, "Médecins de Montpellier."

[20] Hahn, *Anatomy*, chap. 1; Stroup, *Company*; Saunders, *Decline and Reform*, 99–104.

the abbé Bignon, a conscientious advocate of the arts and sciences, to preside over the Academy.[21] Their first two appointments were of the promising botanist, Joseph Pitton de Tournefort, and the chemist and polymath, Guillaume Homberg, who was known in learned circles throughout Europe. Bignon devised ways of encouraging attendance and rewarding accomplishment, and Pontchartrain acceded to the oft-repeated request that academicians be permitted to publish more. When Fontenelle began the eighteenth-century annual reports of the Academy's activities, he specifically thanked the king for supporting the institution even during the difficult wartime years, and his account of activities in 1692 credited Bignon and Pontchartrain with this policy. Finally, the *règlement* of 1699 spoke of continuing, not resuming, pensions and support for the research.[22] Ending financial support would have undermined all these efforts at rehabilitation and jeopardized the broader programs of which the Academy was a component.

For these reasons the suppression of financial support seemed unlikely, and a search for the financial records of the Academy from the final decade of the seventeenth century appeared worthwhile. This search was complicated by the fact that, from 1689 to 1691, Louis XIV reorganized his financial administration. Memoranda from these years reported that the Academy was transferred out of the department of buildings, not that it was suppressed. A document found in the Bibliothèque Nationale recommended small pensions for Homberg and Tournefort, Pontchartrain's earliest appointees. A summary of the state of finances when Michel de Chamillart took over the controller generalship from Pontchartrain in 1699 listed stipends to academicians among the items still to be paid from an account called *gratifications par comptant et autres despenses*. Eighteenth-century records showed that academicians received pensions. Circumstantial evidence pointed to the continuation of financial support by Pontchartrain, but proof was elusive.[23]

Pontchartrain's regime has long been neglected, and historians have deplored the loss of administrative records from this period. Happily the *comptes du trésor royal* from 1683 into the eighteenth century survive in

---

[21] Pellison characterized Bignon as a "personne de mérite, de beaucoup de sçavoir et d'esprit": Leibniz, *Oeuvres*, 1: 289. In the 1720s Réaumur, noting that Bignon had done a lot for the Academy, regretted that "les trésors n'étoient pas entre ses mains": Bertrand, *Académie*, 91–92.

[22] BN MS. Clairambault 566: 251–52; *Histoire*, 2: 132; rules XLVII and XLVIII of the *règlement*; and Fontenelle's statement in *Histoire. . . 1699*, 10–11. Other folios in MS. Clairambault 566 show Bignon justifying expenditure for the Academies and hoping to disburse *jetons* to members of the Académie française on the basis of merit and need, a proposal Pontchartrain rejected (186–93v). Pontchartrain sometimes intervened personally, as when he wrote to the abbé Dangeau of the Académie française, promising to pay for transcriptions and urging Dangeau to "wake up" and finish the work he had promised to do (Clairambault 566: 184v–85). Compare Bignon's and Pontchartrain's task with that of Maupertuis, who fifty years later had to revive the Berlin Academy. He was constrained by insufficient finances, inactive members, and inadequate publications: Brown, *Science and Human Comedy*, 196–97.

[23] BN MS. fr. 7801; BN MS. Clairambault 566: 251–52; BN MS. fr. 22225: 15–23. [Chamillart], "État," 219.

series G[7] of the Archives Nationales. In these documents the payments of pensions and expenses of the Academy are recorded. They prove that even during the troubled 1690s academicians received stipends and that many of the normal expenses of scientific research were reimbursed by the royal treasury. Pontchartrain followed the policies of Colbert and Louvois by pensioning academicians, paying for the costs of research, and maintaining buildings and equipment. Although its members suffered delays in getting paid or reimbursed, the Academy was not excluded from royal funding.

It is difficult to reconstruct the details of the Academy's finances, for the accounts of the royal treasury provide an uncertain and incomplete record. The fourteen boxes containing the accounts for the 1690s hold thousands of pages, each filled with ten to twenty entries about income and expenditure, of which fewer than one hundred items pertain to the Academy.[24] Pensions for members of the Académie des sciences were combined in the records with those for members of the Académie des inscriptions, so that the precise amount destined for either institution is uncertain. Individual academicians sometimes got their pensions sooner than their more patient or less influential colleagues, but usually only a lump sum was recorded for all pensions of the two Academies. Thus the status of several academicians with respect to pensions remains uncertain.[25] Like the buildings accounts, the records from the 1690s obscure the evidence with misspelled names and incorrect or inadequate descriptions. Their organization is inconvenient for researchers on the Academy, for disbursements are recorded in chronological order by date of payment and according to whether they were paid from the *grand* or the *petit comptant*.[26] Academicians were sometimes reimbursed by unknown intermediaries for their expenses, and there is little evidence about the costs of printing their books. Records of the expenses of the Bibliothèque du roi, Imprimerie royale, and Jardin royal offer little more information. As a result, the total sums given here for expenses must be regarded as minimums.

Royal expenditure on the early Academy falls into four general categories: first, pensions, bonuses, relocation subsidies, and wages paid to academicians and their assistants; second, construction and maintenance of the Academy's physical plant; third, the costs of research; and fourth, the costs of publication. In practice, however, the vagaries of the financial record necessitate that some of these categories be conflated and that other classifications be superimposed on the four categories, depending on whether expenditure was direct (intended exclusively for the Academy), or shared (taking the form of disbursements in favor of several royal buildings and institutions including the Academy).

In summarizing the financial data pertinent to the Academy during the 1690s, the following principles have been adopted. Two general categories—pensions and the costs of research—have been defined. The data

---

[24] AN G[7] 892–904. See appendix A.
[25] See fig. 2.1 and appendix C.
[26] Appendix A explains these terms.

related to pensions during the 1690s have been detailed in tables 1 and 2. Data related to the costs of research, including physical plant and publication, are listed in table 5 and analyzed in tables 6 and 7. Table 5 lists all known expenditures related to research, whether they were direct or shared; table 6 separates shared and direct expenditure, organizing them by category; and table 7 analyzes how the research budget was divided between natural philosophy and the mathematical sciences. Engravings and drawings, which are related to publication, and maintenance of the Observatory have been included in research costs, because supervising the drawings and engravings involved scientific vigilance, and the Observatory was one of the Academy's research tools.

Since the records of the Trésor royal on which this study is based are relatively unknown and remain unpublished, they have been described in appendix A, and extracts from them have been transcribed in appendix B.

Assessing Pontchartrain's financial record entails comparing it with those of Colbert and Louvois, but the dates of each minister's fiscal responsibility did not necessarily correspond with his protectorship of the Academy as a whole. Thus Colbert's fiscal regime lasted from 1666 until the end of the fiscal year 1683, even though he died in September 1683. Louvois took control of the Academy in the month Colbert died, but he authorized no expenditure on it until 1684; his fiscal regime began in 1684 and continued until his death in the summer of 1691. Pontchartrain took over the Academy on 25 July 1691, and immediately interested himself in its scientific and financial well-being. He saw to it that academicians quickly received pensions owed to them for 1689 but which Louvois had not paid. Thus, Louvois and Pontchartrain shared fiscal responsibility for the Academy during 1691; Louvois has been credited with all payments made before his death and Pontchartrain with all payments made after Louvois's death. Table 8 compares the finances of the seventeenth-century Academy under its three ministerial protectors, analyzing direct expenditure in three categories— pensions, the Observatory, and research.

A complete record of the Academy's expenses is probably unobtainable, but the data now available form a rich addition to our sources for the history of the early Academy. They clarify Pontchartrain's policies towards the arts and sciences during the difficult 1690s, they refute some common misconceptions, and they permit some comparisons of his protectorship of the Academy with those of Colbert and Louvois.

The following chapters present and explain the data in detail. Chapters 2 and 3 examine the financial and psychological impact of pensions in the double context of ministerial policy and socioeconomic setting, while appendix C explains some of the assumptions behind the discussion of pensions. Chapter 4 assesses institutional morale, taking as evidence attendance (as summarized in table 4), long-range planning, and the style and content of minutes. Chapter 5 treats the Academy's research in the light of its budget. The costs of research reveal hitherto unknown aspects of the Academy's work, suggest a hierarchy of research interests, and clarify the relationship between research and publication during the 1690s. Chapter 6 focuses on one overriding justification for funding the Academy during the 1690s, its utilitarian potential; this draws attention to the use of the scholars to perform propagandistic, cartographic, and clandestine activities and helps explain the Academy's relationship to a sister-society with which it merged in 1699. Finally, in chapter 7 the relative cost of the Academy is weighed and Pontchartrain's effect on the institution is summarized.

Pontchartrain bore two inconsistent responsibilities. As the *contrôleur général* who presided over the royal treasury, he had to fund Louis's wars. As the *secrétaire d'état* who protected the royal academies, he had to revive the Academy of Sciences. It was the Academy's misfortune that the royal coffers were nearly empty during the 1690s, for its recovery depended largely on generous funding.

During the 1690s, therefore, Pontchartrain's protectorship of the Academy was anomalous. On the positive side, he increased the number of academicians and appointed several distinguished savants. He also encouraged academicians to publish, reorganized the Academy in 1699, and continued royal funding of the institution. On the negative side, he refused to authorize resumption of such major research projects as the extension of the meridian, queried even the smallest expenditure on the Academy, delayed the payment of pensions to academicians, and reduced the Academy's budget while founding another learned society which became in some respects the Academy's rival for funds. His overall strategy was to emphasize the Academy's practical expertise and accomplishments while reducing its cost.

As a result, the Academy's health fluctuated from 1691 through 1699. At first it recovered dramatically, in response to renewed funding and the appointment of first-rate scholars to its ranks. But by mid-decade its morale was declining, because of the treasury's failure to pay pensions for several years. Finally, at the end of the decade the Academy began to regain its vitality, after the crown not only improved the institution's finances but also recognized it formally with the *règlement* of 1699.

Pontchartrain defended the Academy within the limits set by military and economic crises. That he did so at all suggests both his commitment to the Colbertian legacy and the crown's expectation that the Academy would be useful. The result was the survival and subtle transformation of an institution which had both a symbolic and a real function in the world of natural philosophy.

II. PENSIONS AND MINISTERIAL POLICIES

The Academy of Sciences had complex responsibilities to its royal founder and to the republic of learning, and its needs were correspondingly manifold. Its members performed scientific research and encouraged others to do the same. They published scientific discoveries, sought the practical applications of theoretical knowledge, and brought *gloire* to the king. To accomplish this multi-faceted program, the Academy required an able and dedicated membership with incomes adequate to free them for scientific research, writing, and discussion. Academicians needed research facilities and supplies, subventions for expeditions, and opportunities to publish their findings quickly and with the appropriate illustrations or tables.

The Academy's health depended on these needs being met. There were three major influences on morale, all dependent on ministerial prerogative—funding, appointments, and authorization to publish—and while generosity in the two latter could sometimes briefly outweigh the deleterious effects of weak funding, it was the royal subvention above all which gave the Academy continuity and distinguished it from its competitors. At its fittest, the academy was genuinely productive, with academicians working harmoniously in teams or reporting individual research to an attentive, respectful, but critical audience, at its twice weekly meetings. At its weakest, the Academy held perfunctory meetings, its laboratory was idle, and members wrote and published little. Ministerial supervision of the Academy was a crucial factor in determining morale, for the ministers appointed academicians, authorized funding, settled disputes, and sometimes urged academicians to initiate or abandon particular research projects.

Throughout the seventeenth century the balance of factors affecting morale and productivity shifted constantly. The best period was during the late 1670s under Colbert; the worst has traditionally been ascribed to Louvois's regime. The decade of the 1690s exhibited dramatic swings in attitude and accomplishment, including a nadir during Louvois's final months, a zenith immediately following Pontchartrain's assumption of the protectorship, a decline as the kingdom felt the worst effects of war and famine, and a revival with the *règlement* of 1699. Pontchartrain resuscitated the Academy by increasing its membership and authorizing an ambitious publication program, but he also diminished and delayed funding, which undermined his other initiatives, as is clear from such indices of morale as long-range planning, attendance, and the style and content of the Academy's minutes.

During the 1690s royal patronage failed to satisfy either the personal or scholarly needs of academicians. Chapters 2 and 3 demonstrate that pensions were inadequate, and chapter 4 shows that morale was injured. Chapter 5 reveals that funding for the costs of research was inadequate, research initiatives were fewer, and publications were as likely to represent old as new research. Yet while the crown never met the needs of academicians during this decade, it nevertheless guaranteed the survival of the institution and laid the basis for its reorganization and development in the eighteenth century.

Appointing and pensioning academicians—acts which at once stimulated their beneficiaries and preserved the Academy—were the prerogatives of each ministerial protector. Together with control of the Academy's research budget, they gave ministers the power to guide the institution and its research. Pontchartrain inherited these prerogatives intact from Colbert and Louvois, but circumstances forced him to modify the policies of his predecessors, with unintended results.

When Colbert decided to pay pensions to members of the Academy, this seemed to him the logical extension of his earlier program of pensions and *gratifications.*[1] In retrospect, his action was as much a break with tradition as an extension of it, for it was a step toward making scientific research a profession. Academicians' pensions were supplemented with lodging, subventions for publications, and research facilities. The intent was to reward and encourage research, but pensions also defined a hierarchy within the Academy and supported ministerial priorities.

Ministerial policies are revealed by examining the size of the Academy, the distribution of pensions within the Academy, and the cost of pensions. Pontchartrain's innovations are best understood in the context of the institution and policies he inherited.

The size of the Academy is summarized in figure 2.1. Colbert founded it in 1666 as an institution with twenty members. During the ensuing years, he appointed sixteen additional academicians; since fifteen academicians died, left, or were excluded, the overall size of the Academy during his protectorship remained relatively stable, with 21.3 members on average. Under Colbert the Academy never had fewer than nineteen members, and at its largest it contained twenty-four members. Louvois inherited an Academy with twenty-one members, and under his protectorship as well the size of the Academy remained generally stable, with a high of twenty-three members, a low of nineteen, and an average of 20.8 members. He appointed six new academicians from 1683 through 1691; again, additions to the Academy were offset by deaths, departures, and exclusions, with

---

[1] The documents refer to "pensions et gratifications," that is, "pensions and bonuses." The Trésor royal distinguished between the two kinds of stipend and called what academicians received "gratifications." But the *règlement* of 1699 described academicians who received stipends on the *estat* of the Academies "pensionnaires," and academicians referred to their stipends as "pensions." I have adopted the Academy's, not the treasury's, terminology.

		Members		No Longer a Member[a]							
		No.	Admitted	No.	Died	No.	Excluded	No.	Lost Pension	Total	Average
Colbert	1666	20	* b	0		0		0		20	
	1667	0		0		0		0		20	
	1668	2	Gallois, Mariotte	0		1	Auzout	0		21	
	1669	2	Blondel, Cassini I	0		0		0		23	
	1670	1	Borelly	0		0		0		24	
	1671	1	Dodart	1	La Chambre	0		0		24	
	1672	1	Roemer	0		0		1	Pivert	23	
	1673	0		1	Gayant	0		0		22	
	1674	1	Du Verney	1	Pecquet	0		1	Richer	20	
	1675	1	Leibniz	2	Frenicle, Roberval	0		1	Niquet	20	
	1676	0		0		0		0		19	
	1677	0		0		0		1	La Voye	20	
	1678	2	La Hire, P. de, Marchant, J.	1	Marchant, N.	0		0		21	
	1679	1	Lannion	0		0		0		20	
	1680	0		0		0		1	Buot (d. 1679?)	20	
	1681	1	Sédileau	0		0		1	Roemer	21	
	1682	3	Le Febvre, Pothenot, Tschirnhaus	1	Picard	0		1	Huygens	21	
	1683	0		0		0		0		21	
	Total	36		7		1		7			21.3
Louvois	1683	1	La Chapelle	0		0		0		22	
	1684	1	Méry	2	Carcavi, Mariotte	0		0		21	
	1685	3	Cusset, Rolle, Thévenot	1	Duclos	0		0		23	
	1686	0		1	Blondel	1	Lannion	0		21	
	1687	0		0		0		0		21	
	1688	1	Varignon	1	Perrault	0		0		21	
	1689	0		1	Borelly	0		0		20	
	1690	0		0		0		1	Cusset	19	
	1691	0		0		0		0		19	
	Total	6		6		1		1			20.8

Period	Year	No.	New members	Departed (Colbert)	No.	Departed (Louvois)	No.	Departed (Pontchartrain)	No.	Total
Pontchar-train	1691	3	Bignon, Homberg, Tournefort		0		0		0	22
	1692	1	Charas	Thévenot	1		0		0	22
	1693	3	La Coudraye, L'Hospital, Morin	Sédileau	1		0		0	24
	1694	4	Boulduc, Cassini II, La Hire, Maraldi	La Chapelle	1		0		0	27
	1695	1	Chazelles		0		0		0	28
	1696	4	Couplet, Guglielmini, Lagny, Sauveur		0	Pothenot[c]	1		0	31
	1697	2	Carré, Fontenelle	Charas	1		0		0	33
	1698	2	Langlade, Tauvry		0		0		0	34
	Total	20[d]			4		1		0	27.6

Summary:	Colbert	Louvois	Pontchartrain
High	24	23	34
Low	19	19	19
No. of years	18	9	8
Average	21.3	20.8	27.6

ᵃ Every academician has been counted as a member of the Academy until death, unless he had been excluded from membership or had permanently lost his pension and stopped participating in meetings. Thus, Huygens and Roemer are not included after 1681, although La Hire and other academicians continued to regard them as colleagues. They were included in the *estat* of 1691 as absent members (BN MS. Clairambault 566: 251-52) in an attempt to persuade Pontchartrain bring them back to Paris. The dates of appointment, death, or exclusion are taken from *IB*, as modified by information in *CdB* and AdS, Reg.

ᵇ The twenty founding members of the Academy were: Auzout, Bourdelin, Buot, Carcavi, Couplet, Duclos, Du Hamel, Frenicle, Gayant, Huygens, La Chambre, La Voye, N. Marchant, Niquet, Pecquet, Perrault, Picard, Pivert, Richer, and Roberval.

ᶜ See appendix C.

ᵈ According to BN MS. Clairambault 566: 251-52, Renaud was an honorary member during the 1690s, but since this marginal note has not been confirmed, he has been excluded from these calculations.

FIG. 2.1. The Size of the Academy of Sciences, 1666-98

the result that the Academy was somewhat smaller when Louvois died than when he took control of it.

Unlike Louvois, Pontchartrain greatly expanded the membership of the Academy. When he assumed control over the Academy in 1691 it had nineteen members (fourteen inherited from Colbert and five from Louvois), but from 1691 to 1698 Pontchartrain packed it with his own men, appointing twenty more academicians. Within eight years, therefore, the Academy grew from its smallest to its largest size (thirty-four members) prior to the reorganization of 1699. Pontchartrain appointed so many new members that the institution grew by almost 80 percent from 1691 to 1698, and it averaged 27.6 members.

The distribution of pensions is summarized in figure 2.2. Most members of the Academy received cash payments in the form of pensions, and some also got bonuses and funds to cover moving expenses. Each of the three ministerial protectors awarded these benefits unequally, according to four categories of academicians: celebrities, regulars, students, and honorary members. Colbert and Louvois pensioned the first three categories, Pontchartrain only the first two. Pensions ranged from three hundred to nine thousand livres a year, and they were based on the merit and reputation of the recipient.[2]

The highest paid academicians were the two celebrities, both of whom were foreign and worked in the mathematical sciences; their pensions were three to four-and-a-half times larger than those of the next highest paid academicians. The higher paid celebrity was the astronomer Jean Dominique Cassini, lured from Italy by a promise to match his Italian income;[3] as a result, his annual pension at the Academy was 50 percent larger than that of the second celebrity, the Dutch mathematician Christiaan Huygens. Unlike other seventeenth-century academicians, both also received funds to pay for the costs of moving to Paris. Support at these levels for foreign savants may well reflect the policy urged on Colbert by Jean Chapelain, who believed that by pensioning scholars from the Dutch and Florentine states, where "letters, like the language, flourish with more brilliance," Colbert would obtain favorable publicity for the king abroad.[4] It also reflects a higher regard for mathematics and astronomy than for other sciences. Cassini and Huygens were powerful academicians who influenced policy and dominated other members of the society. Since the sizes of pensions were no secret, the sums the celebrities received perhaps enhanced their power within the institution.[5] But Louvois and Pontchartrain did not follow

---

[2] BN MS. Clairambault 814: 633r–v, gives this explanation of how pensions were awarded. The manuscript summarizes the history of the Academy under Colbert, describing its founding and its connection with the Bibliothèque du roi; it was probably written to explain the Academy to Bignon when he began to preside over it.

[3] Nicéron, *Hommes illustres*, 7: 309.

[4] Colbert, *Lettres*, 5: 593–94.

[5] Guy Picollet believes that one reason Cassini's treatises were more likely to be printed than Jean Picard's was that Cassini's notoriously high pension earned him more respect. Cassini even got the credit for the work of some of his colleagues, for he formally presented

Colbert's example of importing expensive talent, and thereafter new foreign members received honorary, student, or regular appointment to the Academy.[6]

The regulars and students commanded considerably smaller incomes, ranging from 300 to 2,000 livres a year. Students could expect pensions of 300 to 1,000 livres, while regulars were pensioned at 300 to 2,000 livres. Thus it was possible for a student member of the Academy to receive more than a regular member. Most academicians received either 1,500 livres or between 300 and 1,000 livres and usually collected the same stipend throughout their tenure. But the pensions of a small number changed, sometimes rising, occasionally falling; a few received bonuses from time to time. No formal mechanism existed before 1699 for promoting or demoting academicians from one category to another, or for increasing or diminishing their pensions.[7]

Honorary members received no pensions. Colbert appointed two persons to this position, the German natural philosophers Leibniz and Tschirnhaus. When they visited Paris, the two became acquainted with academicians, and after their departure they kept up a correspondence about scientific matters which was often the topic of discussion at meetings of the Academy.

Colbert's Academy was dominated by regulars who received 1,500 livres. Under Louvois the balance changed, and the numbers of unpaid academicians, regulars who got 1,500 livres, and regulars who got less than 1,500 livres were nearly equal. Furthermore, from Colbert to Louvois, the number and categories of unpaid academicians rose. Until 1675 Colbert pensioned everyone he appointed, and during the eighteen years of his protectorship unpensioned appointments accounted for only 11 percent of the total. Louvois, on the other hand, appointed unpaid members from the start, and during his regime 25 percent of the Academy went without pensions.

Pontchartrain's regime exacerbated these trends. He and Bignon took steps which in the end fundamentally changed the way pensions were distributed. First, in an open break with tradition, they seem to have redefined student membership so that it no longer entailed a pension; con-

---

the *Connoissance des temps* every year: Brice, *Description*, 2: 102. While it is true that the Academy published more works by Cassini than by Picard, nevertheless even some of Cassini's treatises never appeared. He sent many articles to Gallois for publication in the *Journal des sçavans*, but only a few were printed and Gallois kept the rest, of which Cassini had no copy: Cassini, *Anecdotes*, 293.

[6] The foreign members of the seventeenth-century Academy were Cassini, Huygens, Guglielmini, Leibniz, Maraldi, Roemer, and Tschirnhaus. Cassini was naturalized in 1673, four years after becoming an academician: AN PP 151: 68r. Homberg is not included in the list because he was naturalized in 1688, three years before becoming an academician: AN PP 151: 94v. As for the foreigners admitted after 1669, Roemer collected a relatively low stipend in 1672, and from 1677 he received more; Leibniz and Tschirnhaus were admitted as honorary members in 1675 and 1682, Guglielmini as an associate in 1696, and Maraldi as a student in 1694. Although there is no record of naturalization for Maraldi, he was pensioned in 1703 (table 1).

[7] The pensions of Borelly, Bourdelin, Buot, Carcavi, Couplet, N. Marchant, Niquet, Perrault, Picard, Pivert, Pothenot, Richer, Roemer, and Sédileau changed under Colbert and Louvois.

Ministerial Protector	No Pension as Academician	Students		Regulars			Celebrities	Total Pensioned	Total Members	% Pensioned
		300–1,000 lv.	700–1,000 lv.	1,200 lv.	1,500 lv.	2,000 lv.	6,000–9,000 lv.			
Colbert	Lannion	Couplet, C.-A.	Bourdelin	Borelly	Auzout	Buot	Cassini I			
	Le Febvre	La Voye	Pothenot	Buot	Blondel	Carcavi	Huygens			
	Leibniz	Niquet	Roemer	Frenicle	Borelly	Duclos				
	Tschirnhaus	Pivert	Sédileau	Gayant	Bourdelin	La Chambre				
		Richer		Marchant, J.	Buot	Perrault				
				Marchant, N.	Carcavi					
				Pecquet	Dodart					
				Picard	Du Hamel					
				Roemer	Du Verney					
					Gallois					
					La Hire, P. de					
					Marchant, N.					
					Mariotte					
					Perrault					
					Picard					
					Roberval					
					Roemer					
	4	5	4	9	17	5	2	32	36	89

Ministerial Protector	No Pension as Academician	Students		Regulars			Celebrity	Total Pensioned	Total Members	% Pensioned
		400 lv.	300–600 lv.	1,200 lv.	1,500 lv.	2,000 lv.	9,000 lv.			
Louvois	La Chapelle	Rolle	Couplet, C.-A.	Marchant, J.	Blondel	Borelly	Cassini I			
	Lannion		Cusset		Borelly	Duclos				
	Leibniz		Le Febvre		Bourdelin	Perrault				
	Thévenot		Méry		Dodart					
	Tschirnhaus		Pothenot		Du Verney					
	Varignon		Sédileau		Gallois					
					La Hire, P. de					

16

	Ineligible for Pensions		No Pension on Estat	Regulars			Celebrity 9,000 lv.	Total Pensioned	Total Members	% Pensioned
	as Associate, Foreign, or Honorary Members	as Students		300–1,000 lv.	1,200 lv.	1,500 lv.				
	6	1	6	1	7	2	1	18	24	75
Pontchartrain	Bignon, Chazelles, Guglielmini, La Chapelle, Lagny, Langlade, Leibniz, L'Hospital, Thévenot, Tschirnhaus	Carré, Cassini II, Couplet, P.[a], La Hire, G.-P. de, Maraldi, Tauvry	Charas[b], La Coudraye, Morin[b], Sauveur	Boulduc[c], Couplet, C.-A., Le Febvre, Méry, Pothenot, Rolle, Sédileau, Varignon	Marchant, J.	Bourdelin, Dodart, Du Hamel, Du Verney, Fontenelle, Gallois, Homberg, La Hire, P. de, Tournefort	Cassini I			
	10	6	4	8	1	9	1	19	39	49

FIG. 2.2. Academicians and Their Pensions, 1666–98

Sources: IB; CdB; Colbert, Lettres, 5; BN MS. Clairambault 566: 251; AN G[7] 893–904; Leibniz, Lettres, 93–94; and Bernoulli, Briefwechsel, 173, 190–91. I am grateful to Michael S. Mahoney for his advice.

[a] Table 2, note c, assesses the possibility that P. Couplet was pensioned from 1696.

[b] Table 1, fiscal year 1694, note a; table 2, note b; and table 3 summarize the evidence of pensions paid to Charas and Morin off the estat. These pensions are not included in the calculations.

[c] Appendix C discusses the circumstantial evidence in favor of Boulduc's receiving a pension before 1699.

sequently, Rolle enjoyed a de facto promotion to the status of a regular. Second, they increased the number of members traditionally ineligible by definition for pensions. The result was that they more than doubled the number of unpaid academicians, who accounted for 51 percent of the total membership through 1698. The necessity of reviving the Academy during a period of economic crisis led Pontchartrain to transform the old Colbertian policy of pensioning the majority of academicians into a new eighteenth-century rule, according to which the majority of the Academy would not be pensioned.[8]

The overall changes in the patterns of distributing pensions during the thirty-three years spanned by the three ministers can be seen in figure 2.2. These changes were due to the need to economize, which was forced on the ministerial protectors of the Academy as early as the end of Colbert's regime. Once the Observatory was completed in the 1670s, pensions constituted the chief operating expense of the Academy. The contemporaneous Dutch Wars caused royal revenues to falter, and pressures built during the 1680s and 1690s to reduce the Academy's budget. Bowing to exigency, the ministers adopted four expedients affecting members directly: they chose younger members,[9] they diminished the size of pensions awarded to new members, they reduced the number of celebrities,[10] and they appointed more unpaid academicians. These measures ran counter to the generous standards according to which Colbert had originally founded the Academy. Pontchartrain modestly corrected the worst effects of the second measure, but allowed the proportion of pensioned members to fall as the Academy grew in the 1690s.

The change in policy from peacetime generosity to wartime tightfistedness was effective, as is plain from an analysis of the totals budgeted for pensions by each minister, summarized in figure C.2. From 1666 through 1699, Colbert, Louvois, and Pontchartrain budgeted 1,043,575 livres as pensions for academicians, or 30,693 livres a year.[11] Under Colbert the yearly totals for the Academy ranged from 7,950 to 42,800 livres; under Louvois the range was from 7,400 to 27,700 livres; and under Pontchartrain it was from 14,800 to 30,850 livres. Thus Louvois budgeted the least for pensions for the Academy in any one year, Colbert the most. During the 1690s, when the Academy grew dramatically in membership, the cost of pensioning academicians remained low.

Colbert spent an average of 34,140 a year on pensions, compared with Louvois's average of 22,786 and Pontchartrain's of 24,505 livres a year.

---

[8] Twenty of seventy-one members were *pensionnaires* in 1703: BN MS. Clairambault 566: 252v–53r.

[9] Age at entry can be calculated for three-fourths of the academicians admitted from 1666 to 1698. For concerns about the youthfulness of new members see Huygens, *Oeuvres*, 9: 129, 204, 264, 378. Rule XV in the *règlement* of 1699 required that an *élève* be at least twenty-one years old.

[10] Huygens, *Oeuvres*, 8: 456–58, 483–84, 550–53; BN MS. Clairambault 566: 251–52.

[11] This excludes wages paid to the Academy's assistants. See table 8, note a, for those figures.

Under Colbert the average individual pension was 2,029 livres. Thereafter it was always lower. Louvois paid an average pension of 1,437 to his academicians, and under Pontchartrain the average pension was probably 1,498 livres. Colbert, then, budgeted the highest average pensions of the seventeenth century, and Louvois and Pontchartrain the lowest. In fact, as will be seen in chapter 3, academicians collected the smallest incomes under Pontchartrain.

Such averages convey the overall downward trend of pensions in the seventeenth century, but they give a misleading impression of what a normal working member of the Academy could expect to receive. To understand the effects of ministerial policy on regular academicians it is sensible to calculate averages without including the pensions of Cassini and Huygens, because the two celebrities commanded so much more than anyone else that their pensions inflate the overall averages. Excluding Cassini and Huygens, therefore, the average pension for the remaining academicians under Colbert was 1,463 livres, while under Louvois it fell to 1,029 livres, and under Pontchartrain it was probably about 1,027 livres.[12] Thus, during the 1690s, when the cost of living was higher than before, academicians were receiving pensions 30 percent lower than Colbert had paid.

Pontchartrain inherited three goals: to revitalize the Academy, to pension academicians generously so that they could devote themselves to scientific research, and to minimize the costs of the Academy. The third goal was inconsistent with the first two. His solution was to reduce the proportion of academicians who got pensions, but in so doing he altered the pattern of distributing pensions and thereby laid the foundation for the *règlement* of 1699. He also unwittingly changed the social structure of the Academy, which in the eighteenth century had a larger percentage of noble members.[13]

Pontchartrain hoped to revive the institution while limiting its cost. Since the principal cost of the Academy was the budget for pensions, this meant tampering with the traditional structure of the working Academy. He expanded the membership, but contracted the number who received pensions. He tried to overcome the deleterious effects of that contraction by choosing younger candidates, thus extending the potential working lives of academicians, and by appointing large numbers, thereby enlivening meetings and producing more publications. But as will become clear in the next chapter, the treasury could not meet even the modest obligations of Pontchartrain's budget for pensions. For the Academy to revive during the 1690s, therefore, unpensioned or underpensioned academicians needed reserves of enthusiasm and wealth to supplement the minimal incentives provided by the crown.

---

[12] See appendix C.
[13] McClellan, "The Académie. A Statistical Portrait," and Paul, *Science and Immortality.*

III. PENSIONS IN THE LARGER CONTEXT

Pensions functioned as an inducement to work or as a reward for work done on behalf of the Academy. In principle they guaranteed an academician sufficient leisure to conduct scholarly activities. It is important, therefore, to understand their monetary worth to an academician. Unfortunately, specific pensions cannot be correlated with particular styles of life, for situations and tastes varied considerably and many academicians had additional incomes. Indeed, the problem is more general still. Academicians, like professors, physicians, and lawyers, were practitioners of the liberal professions. But although the rise of the liberal professions is one of the hallmarks of social and economic change during the early modern period,[1] the social position, earnings, and standard of living of its practitioners, and especially of savants, have been neglected by historians.[2]

In the absence of either a detailed analysis of the liberal professions under Louis XIV or a complete record of the incomes and expenses of academicians, the position of members of the Academy in contemporary social and economic structures can be gauged only indirectly and imperfectly. It is possible, however, to clarify the relative standing in society of those who received pensions by using several indirect measures: by placing academicians within a broad socioeconomic context; by exploring one individual academician's standard of living on the basis of his testament; by suggesting the source and scope of academicians' other incomes; by assessing the general effects on pensions of the devaluation of the livre at the end of the century; and, finally, by investigating the actual mechanisms for paying pensions during the 1690s. Together these indices hint that pensions were far from adequate, thus jeopardizing the morale of the Academy.

Pensions, as the savants themselves were aware, implied a hierarchy within the Academy. They reflected the kind of work academicians did, their international renown, their access to ministers and the king, and their social origins. Academicians came from varied ranks.[3] The geometer

---

[1] Cipolla, "The Professions," 37.

[2] This omission is now being corrected: see Hahn, "Scientific Careers" and "Scientific Research," and Sturdy, "Chomel," on savants in the seventeenth and eighteenth centuries, and Lehoux, *Le cadre de vie des médecins,* for the way of life of physicians in the seventeenth. The *minutier central* of the AN remains an untapped source of biographical information about members of the early Academy.

[3] It is not clear that the Academy of Sciences tried to abolish differences of rank as the Académie française professed to do: BN MS. Clairambault 566: 195–96.

L'Hospital was a marquis, that is, a nobleman of the sword; Cassini claimed to be, and Tournefort was, a gentleman. Many academicians were clergy, and still others came from those families of lawyers whose sons bought offices in *parlement* or came to Paris in search of a patron. Several represented lower social ranks; as surgeons, apothecaries, or engineers, they had been apprenticed and did not have the university degrees that endowed some bourgeois with status. The Academy of the 1690s encompassed nobility and petty bourgeoisie, and some of its members may have bordered on artisan status; academicians were *honnêtes hommes, honorables hommes, gentilshommes,* but only rarely *écuyers.*[4] Members of the Academy of Inscriptions enjoyed, in contrast, a higher social status and also tended to receive higher pensions, as will be seen later.

Academicians lived and worked in Paris, and the city's social structure in the middle of the seventeenth century has been elucidated by Roland Mousnier's recent analyses, based on notarial records of marriage contracts and inventories after death. He found that practitioners of the liberal professions fell into two distinct categories: some were lawyers, notaries, or physicians with remunerative practices and wealth approaching that of the affluent royal officers of finance; many more existed at the lower end of the scale, with incomes closer to those of day laborers.[5] Mousnier's assessment of social ranks and economic power emphasizes the disparity of earnings within the liberal professions in midcentury, and it remains appropriate in general terms for the late seventeenth century. But to understand the hierarchy of professions during the 1690s it is necessary to turn to other sources, dating from the end of the century.

The capitation tax of 1695 offers such a guide to the kingdom as a whole. Imposed temporarily as a war measure, it was designed to tax individuals more or less according to their means rather than according to their social rank. For that reason it was deeply resented. The *Gazette d'Amsterdam* criticized the tax for ignoring social barriers by lumping together nobles with merchants.[6] Although economic growth and social mobility challenged the old ranking, and another hierarchy based on social classes was emerging, the legal basis for social stratification in seventeenth-century France was not a class or a caste system but the military order, which gave precedence to the nobility of the sword.[7] By defining the tax obligations of subjects in terms of their professions rather than the order they belonged to, and by taxing the royal family and the nobility, the capitation tax challenged the traditional division of society into three orders. Although it did not attempt to gauge real income, it recognized the principle that some people should pay more than others because they were wealthier. Thus the tax clarifies

---

[4] Cassini's claim to be of noble descent was not admitted by eighteenth-century genealogists: Wolf, *Observatoire,* 78–82. For the biographies of academicians, see *DBF, DSB, IB, NBU;* Fontenelle, *Éloges;* Nicéron, *Hommes illustres;* and references in Hahn, *Anatomy,* app. 2.

[5] Mousnier, *Paris au XVIIe siècle,* 232–84.

[6] Pontal, "La Capitation en 1695," 464.

[7] Mousnier, *Social Hierarchies,* chaps. 1–4, 6.

the social hierarchy of occupations, but corresponds only roughly to their earning power.[8]

In the first taxable class were members of the royal family, who were meant to pay a tax of two thousand livres. At the bottom, in the twenty-second class, were day laborers and those who worked with their hands; they owed one livre. Between the first and the twenty-second classes were listed numerous occupations and conditions, some of which applied to members of the Academy. Academicians were not taxed as such, a sign that membership in the Academies was not regarded as a profession in its own right. But some practiced other professions that were enumerated by the tax, especially in the seventeenth class, and their pensions are comparable to the revenues collected by the holders of offices in the fifteenth through twenty-first classes.

The seventeenth class included Parisian doctors, surgeons, and apothecaries, plus professors at the Collège royal, as well as all recipients of wages and pensions from the crown, whether or not they lived in Paris. Members of this class owed 20 livres a year.[9] By comparison, professors of law and physicians in port cities were placed higher, in the sixteenth class, and owed 30 livres, while bourgeois living on annuities in second-order cities were in the fifteenth class and paid 40 livres. Offices in these three classes cost from 200 livres to 60,000 livres, and revenues ranged from 15 livres to 2,400 livres. Medical practitioners and notaries in the smaller cities, barbers, civil engineers, rectors and chancellors of universities, masters of hydrography, and ordinary surgeons in port cities were placed lower, in the eighteenth category, and owed 10 livres. In still lower ranks were university beadles and regents; gentlemen without fiefs or castles; notaries, physicians, surgeons, and apothecaries in very small or enclosed towns; unmarried *valets de chambre* and chambermaids; regimental surgeons-major; married domestic gardeners; and assistants to surgeons, apothecaries, and barbers. Offices in the eighteenth through the twenty-first categories cost from 10 livres to 52,000 livres, and revenues ranged from 10 livres to 1,500 livres.[10]

The capitation tax did not show great regard for the earning power of the liberal professions. Most academicians and their counterparts fell into

---

[8] Bluche and Solnon, *Véritable hiérarchie*; Berton, *L'impôt de la capitation*; Lavisse, *Louis XIV*, 427–28; Mols, *Introduction à la démographie*, 1: 51–52. The text of the capitation tax may be found in Boislisle, *Correspondance des contrôleurs généraux*, 1: 565–74, and in Sourches, *Mémoires*, 4: 502–18.

[9] Under the heading "Dix-septième classe. 20 livres." are listed "Les professeurs du collège royal de Paris, et autres, tant de Paris que de provinces, qui reçoivent gages et pensions." It is not clear whether "autres" refers only to "professeurs" or to those who receive "gages et pensions." In any case, in the terminology of the Trésor royal, academicians received "gratifications," not "pensions" or "gages." On the other hand, academicians were often paid at the same time as the professors royal, and while the latter were usually included in the *chapitre* called "gages du Conseil," sometimes they were listed, along with the Academies, under "gratifications par comptant": AN G[7] 892–904.

[10] For the value of offices and revenues in each category, see Bluche and Solnon, *Véritable hiérarchie*, chap. 2.

the lower third of the spectrum of occupations it defined. The hierarchy the tax reflects for the entire kingdom at the end of the seventeenth century resembles Mousnier's for Paris in midcentury, and in both periods and classifications the incomes commanded by the liberal professions varied widely. The Academy with its unequal, but mainly modest, pensions was a microcosm of the world of the liberal professions.

The relative standing of academicians in the eyes of their patrons or their closest rivals for patronage becomes clearer by comparing their pensions with those of a more select group—physicians, engineers, and scholars paid, like academicians, from the royal treasury during the 1690s. Table 3 summarizes the pensions owed by the crown to members of the Academies of Sciences and Inscriptions, professors at the Collège royal, the staff of the Jardin royal, engineers, physicians, musicians, teachers, and others. These stipends formed three categories: from 6,000 to more than 9,000 livres, from 1,000 to 3,000 livres, and less than 1,000 livres. Physicians got pensions at all levels, engineers only at the second and third, and several professors collected only small amounts. With one exception, members of the Academy of Sciences received pensions in the second and third categories and generally in the lower portion of these categories.

Cassini was the only academician to obtain a pension in the top category. His income was close to that enjoyed by the two first physicians, whose responsibilities and emoluments were enumerated in the *État de la France*.[11] The nine academicians who fell into the second category got somewhat less than professors royal of French law or Greek, leading members of the Academy of Inscriptions, the superintendent of the Jardin royal, or the librarian Thévenot. Seven academicians, along with several professors, engineers, and physicians, collected the mediocre pensions characteristic of the third category.

In general, the crown paid members of the Academy of Sciences less than it did their counterparts. Since the natural philosophers represented lower social ranks than did the humanists, and since no member of the Academy of Inscriptions was paid so little as were most members of the Academy of Sciences, pensions may reflect the social origins of academicians. The crown perhaps appreciated the work of the Academy of Inscriptions, which designed legends and types for the medals and tokens that formed a "metallic history" of the reign, more than that of the Academy of Sciences.[12] If pensions are a reliable guide, the crown had greater esteem for the humanists of the Academy of Inscriptions than for the natural philosophers of the Academy of Sciences.

Information about the incomes of other members of the liberal professions, and especially savants, shows that the smaller pensions available to

---

[11] The *État de la France*, an official publication describing the royal bureaucracy, lists the holders of offices; these were not comparable to membership in the academies or to appointments as professors royal.

[12] Jacquiot, "Pourquoi."

academicians were comparable to what they might have received elsewhere.[13] At the bottom of the scale, François Colletet, editor of the short-lived *Journal des avis*, earned less than two hundred livres in 1680.[14] University professors earned as little as two hundred or as much as two thousand livres.[15] The last of the seventeenth-century editors of the *Journal des sçavans* got fifteen hundred livres a year.[16] Examples abound of scholars living on annuities or small pensions supplemented by room and board. When Bignon began in 1695 to support Antoine Galland as a belletrist, he offered room, board, and a pension of three hundred livres. The marquis de L'Hospital gave Jean Bernoulli the same amount.[17] But even the Academy's assistants for engravings, inventions, chemistry, and dissections earned more, for they were paid four to six hundred livres a year.

A modest pension, room, and board could rescue a struggling, single scholar, but many savants had responsibilities to parents, siblings, wife, or children. A family complicated matters, and Tournefort for one explained that he never married because he did not wish to jeopardize his work.[18] Other academicians did marry and raise children, but we do not know how many were able to adopt the advice Racine gave his son in 1698. He warned that even an annuity of four thousand livres was too little to support a family if one hoped to maintain horses and equipage.[19]

All indices—Mousnier's social ranks, the capitation tax, pensions paid by the crown, and the incomes of savants in general—show that scholars experienced diverse economic circumstances, and that many of the learned

---

[13] On the socioeconomic hierarchy of seventeenth-century Paris and the earning power of various occupations, see Mousnier, *Paris au XVIIe siècle*, 304–306, 340; Crousaz-Crétet, *Paris sous Louis XIV*, 1: 289; Barber, *Bourgeoisie*, 20, 118, 120. D'Avenel, *Histoire des prix*, 4: 3–4, 52, 59; 5: 169, supplies incomes of physicians, surgeons, and provincial schoolmasters, but for the shortcomings of his data and analysis, see Hauser, *Recherches et documents*, 54–55, 75.

[14] An exciseman would have earned seven to eight hundred livres: *Le journal de Colletet*, 51–52, 65.

[15] Table 3 lists *gages et augmentations* awarded to professors at the Collège royal at the end of the century. Brice noted that the rector of the Collège royal had no control over appointments, and that professors royal had the privileges of officers of the Maison du roi: *Description*, 2: 52–55. As the *État de la France* (1663), 2: 432, pointed out, besides the traditional subjects— philosophy, eloquence, and Latin—the Collège royal taught oriental languages, medicine, and mathematics. Just as in the case of the Jardin du roi, then, the crown encouraged reform of learning.

[16] Père Léonard reported at the turn of the century that Cousin's income as editor of the *Journal des sçavans* was eighteen livres "toutes les semaines de la part du libraire . . . , sans les deus cents escus de pension que luy fait donner M. le chancelier." That totaled 1,536 livres: Neveu, "La vie érudite," 466.

[17] Neveu, "La vie érudite," 477–78, on Bignon's support for Galland. L'Hospital began supporting Bernoulli in March 1694: Bernoulli, *Briefwechsel*, 1: 202.

[18] Fontenelle, *Éloge*, eulogy of Tournefort. The records of the treasury sometimes hint at poverty, as was the case with the widow of Fernon, Chantre de la musique du Roy, who collected a pension of one thousand livres as subsistence for herself and her nine children: AN G^7 993. Cournot's notary grandfather raised ten children on only eight hundred livres a year: Lefebvre, *Coming of the French Revolution*, 46.

[19] Racine, *Oeuvres*, 7: 293.

bourgeoisie were relegated to punishing poverty. Natural philosophy was often no more than the avocation of magistrates and physicians, and scholars who lacked a personal fortune or a lucrative occupation sought institutional or individual patronage.

An academician's financial needs depended on many factors, and only the intimate details of daily life recorded in an expense diary might bring to life how an income was spent or how important an academician's pension was to his standard of living. No calculation in the abstract of the cost of living for a Parisian bourgeois in the 1690s can compensate for the lack of such information.[20] But the last will and testament of Denis Dodart (1634–1707) reveals something of how one academician with a family used his money.[21]

In the last two decades of his life, Dodart earned income from several sources. As a physician he had built up a lively private practice among fellow Jansenists, and he also bought the offices of ordinary physician to the king and first physician to the dowager princesse de Conti. To supplement his earned income, Dodart collected annuities from three kinds of investments. First, he held two *contrats de rente sur l'hostel de Ville de Paris*, that is, he received annuities from an investment in the debt of the Hôtel de Ville. On a principal of fourteen thousand livres he earned seven hundred livres; from the other *contrat*, an investment of two thousand livres, he earned one hundred livres in annuities. This kind of investment became increasingly common among the emerging middle class during the seventeenth century; the 5 percent rate of return was high for the period, and such *rentes constituées* were a better investment than rents based on land.[22] But as will be seen, Dodart's shares in the debt of the Hôtel de Ville may have originated not from a desire for sound investments, but from the necessity of making what was very nearly a forced loan to the crown.

Dodart's other investments were most likely chosen out of friendship. As part of the Jansenist circle of intellectuals associated with the duc de Roannez, and as the duke's friend and physician, he bought shares in the company called La nouvelle navigation de la Seine, which was Roannez's brainchild, and in the rights of the creditors of the duke. Other academicians and their associates, including Homberg, Gilles Filleau Des Billettes, and

---

[20] D'Avenel (*Histoire des prix*, 3: 656, 2: 9–10, 303) gives the price of a suit of clothing in 1698 (18 livres) and of houses in 1669 and 1703, and also the rental of houses in the 1660s and 1670s, but there is too little information to support generalizations. Rooms and apartments could be rented, and a genteel woman or a savant in reduced circumstances might rent a room in someone else's apartment for a small sum, as did the widow of Jacques Jaugeon in the 1720s and 1730s: see transcriptions of notarial records in AdS, dossier "Jacques Jaugeon."

[21] Dodart's will is preserved in AdS, dossier "Denis Dodart." Meuvret ("Situation matérielle du clergé," 252) describes the limitations of a testament as evidence of wealth or standard of living.

[22] Braudel and Labrousse, *Histoire*, 2: 343–44; Chéruel, *Dictionnaire historique*, 1065; Vührer, *Histoire de la dette publique*, 116–24; Clamageran, *Histoire de l'impôt*, 2: 667, 3: 26. Dodart left his *rentes* to his children.

L'Hospital, were also connected with these activities, as investors, directors, and inventors.[23]

Dodart also borrowed, without interest, seventeen hundred livres from his friend Louis Morin, another abstemious and pious anatomist, physician, and academician, who lived at the monastery of Saint Victor in Paris. This debt was very much on Dodart's mind, for he wrote "I ceaselessly hope to repay him, if God will preserve my life another two years."[24]

Dodart the investor and debtor is better known as Dodart the Jansenist and pious benefactor of the poor. His modest apparel caused the princesse de Conti, whose first physician he was, to mistake him on the street for a beggar. Yet he employed four or five domestic servants and had inherited a house on the rue Sainte Croix de la Bretonnerie in Paris. He owned a substantial library and furnished his house and his apartments at Versailles, Fontainebleau, and the *hôtel* of the princesse de Conti (on the rue des Poulies) with beds, silver plates, tables, *cabinets*, and a gilded mirror. He also hired carriages, for which he had accumulated a small debt. Dodart had powerful friends to whom he recommended that his son and daughter turn for protection. He named in his will the princesse de Conti and her family; members of the le Peletier and d'Aguesseau families, which supplied *intendants, avocats*, and *procureurs généraux au parlement*, and *ministres* and *conseillers ordinaires d'état*; the abbé Bignon; the *premier président de Paris*; and the *premier président du Grand conseil*. The executor of his will was Brisset, an *avocat* in *parlement*.[25] As his testament shows, Dodart's pension as academician represented less than a third of his income, which by the 1700s derived mainly from purchased offices.

Like Dodart, many academicians depended on income which had no connection to their membership in the Academy. Gallois, La Hire, and Sauveur were professors at the Collège royal, and Couplet (and later his

---

[23] On the duke of Roannez and his circle, see *NBU,* 42: 350, and on the Gouffier family, see 21: 378–79; Colbert, *Lettres,* 1: 210, n. 1. The definitive study is Mesnard's *Pascal et les Roannez;* see 960–73, for the Nouvelle navigation de la Seine. Des Billettes presented several papers to the eighteenth-century Academy on this company; see, for example, his "Description d'une . . . porte d'ecluse."

[24] ". . . je ne dois rien de considérable qu'a Monsieur Morin Docteur en Medecine retiré a Sainct Victor a Paris, mon amy, qui m'a pressé genereusement et sans interest une somme de dix sept cent livres, que j'espere luy païer incessamment, sy Dieu me conserve encore deux ans de vie": AdS, dossier "Denis Dodart," Dodart's will. Dodart died before the two years had elapsed. On Louis Morin de Saint Victor, see Nicéron, *Hommes illustres,* 12: 96–102. Lister visited him in 1698 and described his apartment at the monastery: *Journey,* 132. When Boileau Despréaux had a problem with his throat, Racine sought advice from Dodart, who recommended that Boileau go to his friend Morin, "sans doute le plus habile médecin qui soit dans Paris, et le moins charlatan." Racine, *Oeuvres,* 6: 586–87.

[25] D'Aguesseau (1668–1751) was connected with other academicians, for in 1705 he sent C.-A. Couplet to find a supply of water for the village of Coulanges-la-Vineuse. Like many persons associated with the Academy he was influenced by the Jansenists: *NBU,* 1: 426–34, 12: 176–77; *DBF,* 1: 827–34; Colbert, *Lettres,* 2: 89, n. 2, 548, n. 1. On the street where Dodart had his house the *hôtel* of a *fermier général* was also situated: Hillairet, *Dictionnaire,* 2: 492–93. For the rise of Dodart's descendants, see *DBF,* 11: 417–18.

son) taught mathematics to the pages of the Grande écurie. Du Verney, Tournefort, and Boulduc lectured at the Jardin royal. Truchet and Sauveur taught children in the royal family. Méry was surgeon major at the Invalides and had been associated with the Hôtel-Dieu from about 1681. Some academicians had access to fairly substantial incomes. Gallois, Bignon, and other academicians in orders collected income from sinecures.[26] But others were poor. When Sauveur, for example, petitioned Pontchartrain for his *gages* of six hundred livres and *augmentations* of one hundred livres, earned as professor royal, he pointed out that this was his sole support.[27]

Many academicians must have found their pensions too low to devote themselves exclusively to their research in the Academy, and a petition of 1716 for higher pensions expressed what must have been the sentiments during the 1690s of any academician who received less than 1,200 livres. Even an annuity of 1,500 livres, it asserted, "is inadequate, in Paris, to enable a man to give himself over entirely to the sciences," and 1,000-livre pensions made it impossible to do scholarly work. Nearly a decade later Réaumur described the effects of inadequate pensions. His colleagues had to practice medicine and surgery, sell drugs, or teach mathematics to survive, and one of the anatomists had died impoverished at the Hôtel-Dieu. Under these conditions scientific research for most academicians could hardly be more than an amusement relegated to scarce leisure hours.[28]

The pensions Colbert had paid seemed in retrospect all the more generous[29] in view of three additional problems which must have caused hardships for many academicians. First, the crown seriously delayed payment of pensions. Second, it debased the value of the livre. Third, academicians transformed their accumulated unpaid pensions into investments

---

[26] Sgard et al. (*Dictionnaire des journalistes*, 167–68) say Gallois collected twelve thousand livres a year, but compare Fontenelle's claim in his eulogy of Gallois, in *Éloges*, that during the 1680s Gallois's "abbey was so mediocre that he was obliged to give it up." Clarke, "Abbé Jean-Paul Bignon," 217, 222. On Méry, see *Mémoires*, 10: 324, 656–57, 731. On Sauveur, see n. 27 and chap. 6, n. 12. Tournefort inscribed the names of plants on watercolors in the king's collection and was paid at the rate of two *louis* for each plant: Lister, *Journey*, 62.

[27] Appendix B, document III. Although this petition predates Sauveur's entry into the Academy, it probably exemplifies Sauveur's modest income even after his admission, for he received no pension as an academician. By the late 1690s, however, he was tutoring the dukes of Anjou and Berry.

[28] Bertrand, *Académie*, 86, and 90–91, quotes Réaumur's statement. See also Maupertuis's comments to Bernoulli on the value of seventeenth-century pensions, in Brown, *Science and Human Comedy*, 174. Things may have been worse in the eighteenth century, for the value of the livre continued to fall. In 1716 four academicians (Cassini II, Du Verney, La Hire, and Maraldi) received pensions substantially higher than those of their colleagues. By 1726, when the livre was worth only 45 percent of its value in 1660, the fund for pensions had been raised to thirty-six thousand livres, and Réaumur received one-third of that as his pension. On the effects of inflation for the less well paid eighteenth-century academicians, see Bertrand, "Les Académies d'autrefois" (1867), 752, and *Académie*, 85–107. On the value of the livre tournois, see n. 36, below.

[29] Réaumur believed Colbert's fifteen-hundred-livre pensions were adequate for academicians to live on and guaranteed leisure for scientific work: Bertrand, "Les Académies d'autrefois" (1866): 758–69, and *Académie*, 89–92.

in the debt of the Hôtel de Ville, from which they were entitled to annuities much smaller than their pensions.

During the 1690s, the crown paid pensions to members of the Academies late, sometimes several years after the pensions were due,[30] as can be seen in table 1. This policy began under Louvois, and it continued despite Pontchartrain's efforts to put the Academy on a stronger footing. Perhaps more than any other aspect of royal patronage, the payment of pensions reveals the constraints within which Pontchartrain balanced his responsibilities to scholarship against those to the treasury. One of his first acts on becoming protector of the Academy in 1691 was to authorize payment of the pensions still due from 1689. In January 1690 Louvois had paid only one-third of the *estat* for that year, and academicians had worked throughout 1690 and most of 1691 without receiving any further installments. Thus, when Pontchartrain obtained approval for payment of the remainder, only two months after becoming responsible for the Academy, he demonstrated his good faith to academicians. In 1692 he completed payment of the pensions academicians had earned during Louvois's protectorship, and in 1693 academicians received their pensions for 1691.[31] But after this auspicious beginning, the increasing deficits of the treasury prevented Pontchartrain from paying pensions.

No academician received before 1695 a pension for work performed in 1692. In 1695 and 1696 academicians petitioned for their back pay from 1692 through 1694, with inconsistent results. When Cassini applied in January 1696 for his pensions of the past four years, Pontchartrain authorized cash payment of only one year's worth of back pay.[32] Even when payment was authorized, however, there was no guarantee that funds existed, and one petition bears the note "Pay it after tomorrow if possible."[33] Thus, in February 1695, Charpentier, a member of the Academy of Inscriptions, won the right to be included on the *état de distribution*, but he waited several months to realize the benefits of inclusion.[34] The crown settled academicians' claims for 1692 through 1694 at the end of 1696, but academicians could not count on regular payment of pensions thereafter. Their claims for pensions earned in 1695 were not settled until 1698 or 1699. The pensions due for 1696, 1697, 1698, and 1699 were paid more promptly.[35] From January 1690 until January 1700, every academician en-

---

[30] The crown was late in paying many persons, and academicians got their pensions faster than some of the others; AN G[7] 902, for example, lists payments in fiscal year 1698 of several pensions due to non-academicians from 1692.

[31] Table 1, fiscal years 1690, 1691, and 1692.

[32] Appendix B, document V.

[33] AN G[7] 994, 29 Apr. 1695. This document did not concern the Academies, but is symptomatic of the general problem.

[34] AN G[7] 993, 22 February 1695. Appendix A discusses the *état de distribution* and documents III and V in appendix B are requests to be included on it.

[35] See table 1.

titled to a pension from the crown was injured by the treasury's failure to honor its obligations punctually.

The harsh effects of these delays were exacerbated by the debasement of the currency during the 1690s. After 1689 the livre began to fall in value. From being worth 8.33 grammes of fine silver (as established by law in the 1640s), it fell to 7.56 between 1689 and 1691, rose slightly to 7.80 from 1691 to 1693, fell again to 6.93 from 1693 to 1699, and dropped lower still to 5.31 from 1699 to 1709.[36] This had the effect of raising prices, which were already unstable because of the famine of 1693 and 1694.[37]

When the crown did fulfill its obligations to academicians for 1692 through 1694, that settlement was in some sense a fiction. While the treasury formally discharged its debt on paper, only a small percentage of the money due was realized as cash by academicians. The reason for this paradox is that many academicians converted their pensions into investments in the debt of the Hôtel de Ville, from which they were entitled to annuities. The crown frequently offered such investments in hopes of raising cash for the treasury. In 1692 and 1693 it invited anyone, whether French or foreign, husband or wife, adult or child, to invest in these *rentes constituées*, which were available in two forms: the first could be inherited, and they paid an annual sum equal to one-eighteenth of the principal; the second, *rentes viagères*, could not be inherited but paid higher rates of interest on a sliding scale, depending on the age of the investor.[38]

The transformation by academicians of their pensions into annuities was an ambiguous solution to their plight. On the one hand, that transaction resembled a forced loan to the crown and resulted, in the short run, in a much reduced income for the academician. For example, if a scholar entitled to a pension of 1,500 livres a year invested three years' worth of accumulated pensions, or 4,500 livres, in the debt of the Hôtel de Ville, he would collect, at best, 500 livres from a *rente viagère*.

Furthermore, the crown's treatment of individual academicians was inconsistent. Pontchartrain sometimes sweetened the arrangement with an academician by increasing slightly the principal invested. Some academicians who took *rentes viagères* were assigned an annuity worth one-ninth of the principal, regardless of their age. For Gallois, who at sixty-three ought to have collected one-seventh of the principal, this was a disadvantage, while for the abbé Paul Tallemant, who was forty-three and entitled to one-tenth of the principal annually, it was an advantage. As with the award of pensions themselves, the crown did not treat academicians equally.[39]

---

[36] In 1726 it was worth 4.5 grammes. Braudel and Labrousse, *Histoire*, 2: 345–46.

[37] Lavisse, *Louis XIV*, 2: 422.

[38] Louis XIV, *Edit du Roy, portant création de douze cens mille livres de rente*, and *Edit du Roy, portant création de rentes viagères*.

[39] See appendix B, documents I, II, IV, and VI, for some of the arrangements made with members of the two Academies.

On the other hand, it was better to receive a small annuity than nothing at all. Moreover, so long as the academician had not invested in a *rente viagère*, the principal was not forfeited and could in theory be collected later from the treasury, thereby acquitting the crown of further responsibility to pay annuities. Finally, living on an annuity was prestigious. Furetière reported the received opinion that a *bon bourgeois* "lives on his annuities," and "has no job or responsibility," or, that it is a poor man indeed "who has neither annuities nor a smallholding" and "lives from the labor of his hands."[40] But the fact that academicians did not abandon the Academy in order to live on their annuities shows that membership in the Academy, which entailed working for an eventual pension, was more advantageous than retiring on a small annuity.

According to the abbé Tallemant, by the spring of 1695 most members of the Academy of Sciences had decided to invest their back pay in the debt of the Hôtel de Ville, and Tallemant believed that most members of the Academy of Inscriptions would also choose to do so, if given the opportunity. Calling the arrangement an "invention" and a "kindness," Tallemant petitioned for it himself. But as the documents show, many academicians, including Tallemant himself, took this step only when their appeals for cash payments were unsuccessful. Here, too, some academicians were favored over others, and members of the Academy of Inscriptions were more likely than members of the Academy of Sciences to obtain the full cash value of their pensions for one or two of the years when scarcely any pensions were being paid.[41]

There is no question that the last decade of the seventeenth century was a period of privation for academicians. They endured long intervals without receiving pensions for their work and then had to settle for annuities instead of the full sums owed them. Even those with additional income felt the hardships of the 1690s, for the exceptional measures of taxation touched even the incomes of the clergy.[42] If ever it was fair to mock Louis's savants for having been bought cheaply,[43] it was at the end of the century.

Not surprisingly, as the next chapter will show, attendance and morale were low at the Academy of Sciences during the 1690s.[44] But once the War of the League of Augsburg came to an end in 1697, Pontchartrain tried again to restore morale. In 1698 he obtained approval for the payment of

---

[40] Furetière, *Dictionnaire universel*, "rente."

[41] Appendix B, document IV. A survey of AN G⁷ 991–97 shows that in the Academy of Sciences Cassini got an authorization for a cash payment for 1692, while in the Academy of Inscriptions Tourreil got one for 1695, Tallemant for 1692, Racine and Boileau Despréaux for 1693, and Renaudot and Charpentier for years not specified. These details are confirmed and amplified by AN G⁷ 894–99.

[42] Appendix B, document IV.

[43] As the historiographer Duclos did in his *Mémoires secrets sur le règne de Louis XIV*, according to Peignot, *Documents authentiques*, 120–22.

[44] The Academy of Inscriptions seems to have suffered less in these respects: Jacquiot, "Pourquoi."

pensions for 1696 and 1697, charging both *estats* to fiscal year 1697, and the *estats* for pensions earned in 1698 and 1699 were presented even before the working year was concluded.[45] Pensions were paid more promptly in the final years of the century. These remedies, with the *règlement* of 1699, were intended to put the Academy on a new and more certain foundation.

Pensions were initially conceived by Colbert as reflecting a hierarchy of merit and reputation and providing support for full members. From the 1680s pensions were reduced to rewards and incentives. During the 1690s these stipends remained less generous than they had been under Colbert, and a desperate economy, caused by bad harvests, war, debased currency, and higher prices, made it impossible for the crown to pay academicians at all in some years. At the end of the century, even with lodging provided, being a pensioned member of the Academy of Sciences would not have provided a comfortable income for any but Cassini. Whether viewed as rewards or incentives, pensions were inadequate and too unreliable during the 1690s to induce most members to work vigorously in the Academy. By the middle of the decade, once the momentum of Pontchartrain's early ministerial initiatives subsided and the material rewards of belonging to the Academy were further reduced, the morale of the institution was endangered.

---

[45] See table 1 and [Chamillart], "État."

PLATE 1. Louis Phélypeaux de Pontchartrain, Secretary of State and Controller General of Finances. Portrait by Voligny during the 1690s.

IV. INDICES OF DECLINING MORALE

Pensions were reserved for working members of the Academy, who were expected to attend meetings. These were held twice a week, on Wednesday and Saturday afternoons, except for a six- to eight-week holiday from the beginning of September until mid-October or early November. Although Wednesdays were reserved for the mathematical sciences and Saturdays for natural philosophy, pensioned members were required to attend on both days,[1] and scientific papers belonging properly to the Saturday meetings sometimes were presented on Wednesday and vice versa. Because many members had genuinely interdisciplinary interests, this dual attendance usually was agreeable.

The twice-weekly meetings were the focal point of the institution. At these meetings academicians argued new hypotheses, presented solutions to problems, reviewed books, reported findings, observed demonstrations of experiments, and examined the notebooks from the chemical laboratory. They also planned for the future. Planning enhanced the corporate character of the society and the power of the ministers, for academicians revised one another's proposals and ministers authorized work or submitted objections and counterproposals. By playing a role in planning, ministers stayed familiar with the research they funded and could adjust its direction. Whether research projects were collaborative or individual, meetings provided the principal forum for the Academy. Attendance was therefore essential for any academician who wished to shape the organization, direct its research, or take advantage of some of the benefits this scientific society offered.

The morale of the early Academy depended, therefore, on its meetings, for academicians were not only independent researchers, but also formed units which planned, did research, debated, and even lived together in the Academy's premises. Morale fluctuated between the founding and the reorganization of the Academy, but it declined in the 1690s, as is demon-

---

[1] The Academy's weekly and annual schedule may be inferred from the seventeenth-century minutes, AdS, Reg., 1–18, and from Claude Bourdelin's notebooks of chemical experiments: AdS, Cartons 1666–1793, nos. 1–3 and BN MS. n. a. fr. 5133–49. Francis Vernon reported the annual vacations: Oldenburg, *Correspondence*, 6: 293; 7: 141, 271. All the Parisian Academies took their vacations at the same time. A Wednesday or Saturday meeting might be canceled, and exceptional meetings were sometimes held. From the beginning, that is, on 22 Dec. 1666, it was agreed that because of the close connections among the sciences all academicians should attend both meetings: AdS, Reg., 1: 1. Carcavi and Huygens told Vernon that academicians "did not make inquiries into any one subject in particular, butt every one tooke unto his examination what suited best with his owne Fancy & genius": Oldenburg, *Correspondence*, 5: 498; see also Brown, *Scientific Organizations*, 159.

strated by the style and contents of the minutes, the infrequency of long-range planning, and absenteeism.

The minutes of meetings during the 1660s convey vigor, imagination, and serious dialogue among colleagues. But by the early 1670s, minutes were not kept and projects like the natural history of plants became disorganized.[2] Once the habit of keeping minutes was reestablished in 1675 and 1676, it never lapsed again, but the style and content of these records shifted subtly over the rest of the century. During the 1660s, 1670s, and 1680s, the minutes are rich in detail. They reproduce short papers, long monographs, and correspondence; they report visitors and experiments; they divulge bitter rivalries; and they reveal spirited discussions. But in the mid-1680s there is a change in content and tone, for discussions lost their sense of direction and spirit. From 1686 through 1691 the minutes support Fontenelle's claim that, before Homberg was admitted in 1691, academicians were at a loss for ways to fill their two-hour meetings. One of Homberg's virtues was his ingenuity in demonstrating experiments or presenting papers, and for a while interest in the biweekly meetings was rekindled.[3] But even with Homberg's contributions, the minutes of the 1690s seem laconic, for they contain fewer scholarly papers and do not hint at lively debates of the sort that had characterized the Academy's meetings in earlier times. From the mid-1690s the minutes show that even Homberg's energies flagged.

Long-range planning is another index of institutional vigor. Frequent, regular, and detailed long-range proposals mark a vigorous institution; their absence betrays slumps in the Academy's morale. There were twenty-five proposals for long-term botanical research presented from 1667 to 1699, only three of which were made during the 1690s. In the early period academicians had planned a year's work and developed several projects, such as the natural history of plants and the comparative anatomy of animals, which continued for decades. In the 1660s, before experimental facilities were ready and as academicians became accustomed to corporate research, detailed proposals for research were offered regularly. In the 1670s and 1680s academicians often filed new proposals as they analyzed the progress of their research or responded to ministerial inquiries or initiatives. In contrast, of the three proposals presented in the 1690s, only the last, in February 1699, planned a year's work, and it followed the reorganization.[4]

---

[2] When Du Hamel joined a diplomatic mission to England in the late 1660s, Gallois briefly took his place and was pensioned for this work in 1668 and 1669: *CdB*, 1: 283, 299–300, 377–79. But even after returning to Paris, Du Hamel did not immediately become active again in the Academy: Oldenburg, *Correspondence*, 7: 33–34. As a result, the Academy's business fell into disorder and no minutes were kept from 1670 through 1674.

[3] Fontenelle, *Éloges*, eulogy of Homberg. On the decline of the Academy under Louvois, see Saunders, *Decline and Reform*, and Stroup, *Company*.

[4] Memoirs about long-range planning for botanical research from 1666 to 1699 appear in AdS, Reg., 1: 1–4 (31 Dec. 1666); 22–38 (15 Jan. 1667); 248–51 (14, 21, 28 Jan. 1668); 254 (4 Feb. 1668); 4: 58r–63r (16 June 1668); 6: 180–82, 183–88 (23, 30 Nov. 1669); 8: 5–7v (23 Feb. 1674); 3–5r (23 Jan. 1675); 62 (28 Aug. 1675); 114–16v (26 May 1677); 117r, 120v–27v, 134

Absences from meetings also reveal declining morale. Indeed, the very fact that Pontchartrain and Bignon required attendance to be recorded in the mid-1690s suggests that the problem was serious. In doing so they were trying to encourage industry among academicians, but a side-effect of documenting attendance is an improved historical record. Until the mid-1690s the minutes list only the major participants in meetings: academicians who read papers, discussed issues, or presented experiments, plus visitors. The minutes were primarily intended to indicate work accomplished and ideas exchanged, and they only coincidentally, and hence haphazardly, recorded attendance. Because of Pontchartrain's and Bignon's requirement, a fuller analysis of attendance is possible for the period immediately before the reorganization.

An informal survey of the minutes of the seventeenth-century Academy suggests that many academicians participated in both Wednesday and Saturday meetings, with a core of members dominating the Academy by virtue of their steadfast attendance. But a systematic analysis of selected meetings in the years immediately before the reorganization reveals how small the working Academy was under Pontchartrain, at the time when it had the largest number of members. The dominant core under Colbert and Louvois represented about two-thirds of the membership of the Academy, but under Pontchartrain it accounted for closer to one-third of the membership. Because institutional vitality depended on participation in meetings, these data demonstrate the malaise that weakened the Academy at the end of the century.

Although it may appear that the Academy was more industrious under Pontchartrain because it met more frequently, absenteeism was common.[5] A survey of the eighty-two meetings at which botany was discussed during the period from 5 January 1695 until 7 January 1699 suggests just how few members could be counted on to contribute (table 4). Seven of the thirty-six members—Bignon, Chazelles, Guglielmini, Langlade, La Coudraye, Leibniz, and Tschirnhaus—never attended at all. As president, Big-

---

(2, 30 June 1677); 135 (21, 28 July 1677); 173r–78r (18 May 1678); 189–91r (16, 23 Nov. 1678); 203r–v (24 May 1679); 10: 57v (8 Jan. 1681); 96v (22 Apr. 1682); 124r–v (24 Mar. 1683); 11: 1 (23 June 1683). In subsequent years Dodart once gave instructions to Jesuits about to depart for China (11: 115r, 20 Dec. 1684) and once reiterated his earlier plans (11: 144v, 21 Nov. 1685). These are minor and unlike other proposals. La Chapelle's speech of January 1686 stimulated renewed planning: 11: 155v–56r, 157r–58r (30 Jan. 1686); 162v, 163r–64v, 166v, 167r (2, 13, 27 Mar. 1686); 168r–69r (3, 17 Apr. 1686). When Homberg and Tournefort entered the Academy, once again plans for botanical research were presented: 13: 70r (28 Nov. 1691); 71r–72r (12, 19 Dec. 1691). The reorganization of 1699 stimulated the final long-range planning of the century: 18: 131r–46r (28 Feb. 1699).

[5] Saunders, *Decline and Reform*, 41–42, 47, points out that in the twenty-four years from 1676 through 1699 the Academy met 1,679 times. Breaking that period down into three eight-year segments, corresponding with the three protectorships, Saunders found that under Colbert and Louvois the Academy held 1,100 meetings, 550 under each protector. Under Pontchartrain, therefore, the Academy met more often, for a total of 579 meetings in eight years. In the absence of yearly figures for the 1690s, it is not clear whether the additional meetings took place before or after the reorganization, or were evenly spread over the eight years from 1692 through 1699.

non did not need to attend meetings, and usually he simply reviewed the minutes.[6] Chazelles, Guglielmini, Langlade, Leibniz, and Tschirnhaus were external, associate, or honorary members, not expected to attend but to share their work with the Academy through correspondence.[7] La Coudraye, as a geometer, was subject to the same requirement as any regular and had no such excuse. Every academician was absent at least once. Bourdelin and Du Verney represent the two extremes among the natural philosophers: the former had the best record, missing only one meeting despite his infirmity and age, while the latter missed sixty-seven of the meetings surveyed.

The extent and effects of absenteeism can be appreciated by grouping academicians according to their title, which was based on their research fields or responsibilities (botanist, chemist, mechanician, secretary, and so on); their pensionable status in the Academy; or the ministers who appointed them. Comparisons along these lines reveal patterns of absenteeism that are sometimes surprising.

Most of the eighty-two meetings surveyed took place on Saturdays, which were set aside for natural philosophy, and while most academicians were required to attend, the botanists, chemists, and anatomists had a vested interest in doing so, for this was the opportunity for them to discuss the subjects closest to their own concerns. We would expect the natural philosophers of the Academy, therefore, to have the lowest level of absenteeism, with astronomers and geometers understandably absent at higher rates.

But botanists as a group were absent from 37.2 percent of meetings at which botanical research was reported. Dodart, the director of the natural history of plants, who had revived the project in the 1670s when he was a new member, missed 29.3 percent of the meetings. As physician to the dowager princesse de Conti he was frequently absent from Paris, and after the *règlement* of 1699 he needed a special dispensation to receive his pension in view of his absences.[8] Tournefort, chosen to revitalize botanical research, missed more than half of these meetings. Jean Marchant, who missed only 8.5 percent of the meetings, did the most mechanical work, describing plants for the natural history, and Morin de Toulon, who was called a botanist, actually worked on mineralogy and was absent from 53.7 percent of the meetings. The four chemists who were regulars—Bourdelin, Homberg, Charas, and Boulduc—also had a responsibility for botanical research, but they were absent from 29.1 percent of meetings where botany was discussed. The anatomists—Du Verney, Méry, the student Tauvry, and Du Hamel from 1697—missed 33.5 percent of the meetings. Taken together,

---

[6] Saunders, *Decline and Reform*, 147–48.

[7] Chazelles, for example, contributed to the Academy's work by sending the coordinates of Mediterranean cities to Cassini (see chap. 6), and Leibniz not only kept up with the Academy's affairs through correspondence but also forwarded a manuscript on dynamics to be reviewed by academicians: Leibniz, *Lettres, Oeuvres*; Costabel, *Leibniz and Dynamics*.

[8] AdS, Reg., 18: 115r; in 1691, according to Pellison, Dodart traveled to Versailles two or three times a week to discharge this responsibility: Leibniz, *Oeuvres*, 1: 217.

the natural philosophers missed one-third of the meetings, which suggests that botanical and anatomical research, formerly conducted as lively if frustrating team projects, had become fragmented.

Among academicians whose principal interest was not natural philosophy, geometers and mechanicians not surprisingly had a higher rate of absenteeism than the natural philosophers. The two mechanicians, Couplet and his son, missed 50.7 percent of the meetings, and the six geometers who were regulars or students missed 43.1 percent, or 33.1 percent if Pothenot was excluded in 1696. But the six astronomers were more faithful even than the botanists and anatomists, missing only 27.2 percent of the meetings where botany was discussed. Taken as a group, these academicians missed 36.8 percent of the meetings; if Pothenot was excluded in 1696, they missed 32.7 percent of the meetings and thus had fewer absences than the natural philosophers.

The Academy's secretaries ought to have had few absences, for they were responsible to some extent for the smooth functioning of the Academy and the maintenance of its records. One of their tasks, for example, was to see to it that papers or proposals presented at meetings were copied into the minutes. During almost the whole of the period before the reorganization, the position of secretary was occupied by Du Hamel. But in the 1690s Du Hamel's commitment to the institution changed as he enlarged the responsibilities of the secretary by embarking on a history of the Academy;[9] in 1697 Fontenelle was formally named permanent secretary to replace Du Hamel, who remained in the Academy as an anatomist. Discontinuity in such an important post could have affected the business and records of the institution if it had been coupled with habitual absence, but the two secretaries had the best record of attendance of any group in the Academy. Together they were absent from only 11.1 percent of the meetings which they were eligible to attend as secretary.

Taking academicians according to their pensionable status shows that among those who were required to attend, except for Cassini, higher status correlates with higher absenteeism. Cassini missed 28 percent of the meetings, regulars 34.1 percent, and students 26.4 percent; pensioned members missed 28.4 percent of the meetings, unpensioned members missed 47.1 percent. Among academicians who were not required to attend, only two—L'Hospital and Lagny—came occasionally, but they missed 95.9 percent of the meetings. Hence, while a pension was no guarantee that an academician would come to meetings faithfully, it encouraged attendance.

From the standpoint of ministerial appointment, those whose membership predated Pontchartrain showed more loyalty. Colbert's eleven academicians missed 33.4 percent of the meetings and Louvois's three missed only 19.5 percent, while Pontchartrain's academicians, excluding those who were not required to attend, missed 44.7 percent of the meetings surveyed. Overall, the twenty-eight academicians expected to attend were absent

---

[9] Leibniz, *Lettres*, 93.

from 36.8 percent of the eighty-two meetings at which botany was discussed.

Without comparative figures for earlier or later periods, or for the meetings at which botany was not discussed, the full significance of this high absentee rate is uncertain. But since eleven of twenty-eight, or 39.3 percent, of the academicians who were required to attend meetings missed more than 35 percent of these meetings, and only seven of twenty-eight, or 25 percent, missed less than 10 percent of the meetings, it seems clear that at the end of the century something was amiss. Thus, Martin Lister, an English natural philosopher who visited Paris in 1698, did not exaggerate the small size of the working Academy when he estimated its numbers at a mere twelve or sixteen, even though at that time twenty-eight members were expected to attend meetings.[10] The Academy had become for many members merely a prestigious affiliation and, for some, the source of an irregularly paid stipend; it was no longer their principal forum for stimulating exchange of opinions or data.

The reasons for this lack of commitment were varied. As the percentage of unpensioned members rose, so did absenteeism. Likewise, untimely payment of pensions discouraged attendance. Factions and envy were perpetual problems,[11] and some members did not relish the interdisciplinary meetings. The astronomers were often abroad. Disenchanted with their collaborative research, some botanist academicians elected to work independently, and Tournefort preferred herborizing to the kind of research his fellow botanists did. But the very fact that such individual preferences overrode the requirement to attend suggests the weakness of the Academy as an institution during the latter part of the 1690s.

The old Academy, designed by Colbert and dominated by his academicians until the 1690s, faltered with the entrance of new personnel who brought fresh research interests but lacked loyalty to the established institution, its former members, and unresolved research projects. The deaths of several stalwarts in the 1680s and 1690s, the decline under Louvois, and the infusion of so many additional members by Pontchartrain had the effect of challenging the Academy's traditions. Perhaps Du Hamel was prompted to write his history of the Academy to commemorate the happier years under Colbert or as an antidote to change and rootlessness, but in any case, the forces which alienated academicians from the institution in the 1690s were too strong. As evidenced by the style and content of minutes, the decrease in long-range planning for botanical research, and irregular attendance at meetings, during Pontchartrain's protectorship meetings became less interesting, morale diminished further, and the Academy stagnated. The accuracy of this conclusion is reinforced by an examination of the research program of the Academy during the 1690s.

---

[10] Lister, *Journey*, 80. Fewer attended than collected pensions in 1698; see fig. C.2.

[11] Pellison wrote to Leibniz that the Academy was "assez sauvage" and that "parmi des personnes de grand mérite, il y en a quelques autres sujettes à ces sottes jalousies dont les gens nourris à l'ombre et hors du commerce du monde sont quelque fois plus prévenus que les autres": Leibniz, *Oeuvres*, 1: 289.

V. THE RESEARCH PROGRAM AND ITS COSTS

The Research. The Academy passed through four distinct phases during the 1690s. The first lasted until Louvois's death in July 1691 and was characterized by relative inactivity: fewer papers were presented at meetings and the laboratory was idle. Nothing at all was published in 1690. The second began once Pontchartrain and Bignon took over: long and theoretical papers were read more often than in the immediately preceding years and several books appeared. In 1692 and 1693, academicians published eighty-five articles, almost as many as they had published in the *Journal des sçavans* since 1666, although many of the new articles recorded activities of prior years.[1] This revival continued until late in 1695, after which energies flagged once again, and the third phase began, in which official publications declined and the number of presentations at meetings fell each year from 1695 through 1698.[2] The adoption in January 1699 of formal regulations governing the Academy launched the fourth phase, which renewed corporate energy and inaugurated annual publication of the *Histoire et mémoires.*

Throughout the 1690s research continued along lines already established under Colbert and Louvois. Few new theoretical impulses motivated the natural philosophers or astronomers. What natural historical novelties caught the attention of the Academy were sent by correspondents and inspired only brief descriptive memoirs. Only Tournefort's system of classification, Morin's study of minerals, and the work of L'Hospital and Varignon on infinitesimal calculus represented a new direction for academic research.

Despite declining morale, some of the research the academicians pursued shows continuity over several years. Such long-term projects included measurement of rainfall, analysis of the movement of liquids, observations of eclipses and of Jupiter's satellites, studies of eastern and western calendars, preparation of the natural history of plants, studies of the elasticity and pressure of air, experiments with the air pump, work on poisons and

---

[1] On the difficulty of sustaining this level of publication, see Lister, *Journey,* 80, 97, and "Publications," in this chapter. While it is true that these papers had not as a rule been presented at meetings before 1690s, as Saunders (*Decline and Reform,* 168–70) points out, many built on earlier research. But see n. 33, below. The monthly articles which appeared in 1692 and 1693 as official publications of the Academy were reissued in vol. 10 of *Mémoires,* along with the articles academicians published in the *Journal des sçavans.*

[2] Saunders, *Decline and Reform,* 169, on the number of scientific presentations each year. According to these figures, academicians averaged 166 a year under Colbert, 159 a year under Louvois, and 160 a year under Pontchartrain.

their antidotes, and studies of comparative anatomy, especially the anatomy of the fetus. Mathematicians considered theoretical problems, such as methods of determining the roots of irrational cubes, and also problems of mechanics and hydrostatics. Members received letters containing observations of eclipses and details of natural curiosities. Sometimes academicians reviewed new literature, as when they discussed Swammerdam's works. Two Parisian inventors who were not members, one an instrument-maker, the other an engineer, submitted designs of inventions. The student members contributed by inventing machines, observing fixed stars, calculating solar and lunar diameters, and traveling abroad to collect astronomical and natural historical data. Finally, Du Hamel worked throughout most of the decade on his history of the Academy.

Publications included collected papers; the astronomical observations of Jesuit missionaries in the Far East, as analyzed by La Hire and Cassini; treatises by Cassini, Lagny, La Hire, L'Hospital, Rolle, and Tournefort; La Hire's edition of the manuscripts of deceased colleagues; and Du Hamel's history. Some of these works had begun under Colbert and Louvois, and Pontchartrain made it possible for academicians to achieve their hope of many years' standing to publish them.[3]

Reduced Costs. Given the temporary revival of research in the early 1690s, as compared with the late 1680s, and the dramatic increase in membership of the Academy, costs ought also to have climbed in the 1690s. But depleted royal coffers actually reduced the costs of research. As table 7 shows, 36,932 livres of direct expenditure by Pontchartrain on the Academy's research from 1691 through 1699 have been identified, although of that, only 19,218 livres were spent from 1691 to 1698. In addition, there were indirect expenditures, such as the costs of publishing the books and articles of academicians at the Imprimerie royale, which could be high. The plates alone for Tournefort's *Élemens de botanique* cost twelve thousand livres.[4] Finally, the Academy benefited from shared expenditure, that is, from some part of 69,037 livres spent for rent and maintenance of the Bibliothèque du roi, the Jardin royal, and other royal buildings, as well as from 11,850 livres spent on illustrations of plants for the Bibliothèque du roi (table 6). Excluding publication costs, about which all too little is known, direct expenditure on the Observatory and the Academy's research averaged 4,103 livres a year from 1691 to 1699, compared with 52,045 livres a year under Colbert and 8,170 livres under Louvois (table 8). But the direct expenses identified to date probably represent only part of the money spent for academic

---

[3] For the Academy's activities during the 1690s, see *Historia*, 301–524; *Histoire*, 2: 133–344; *Mémoires*, 10: 1–448, 734–44; and Leibniz, *Lettres*, 79–118, and passim. See Saunders, *Decline and Reform*, 66–91, on attempts to convince Louvois to publish more; Stroup, *Company*, chap. 15; and Bernard, *Histoire de l'Imprimerie royale*, "Catalogue chronologique des éditions de l'Imprimerie royale du Louvre," 138–54, for what the Imprimerie royale printed for the Academy from 1666 to 1699. On the work at the Observatory, see Wolf, *Observatoire*, 146, 174–75, 206.

[4] Laissus and Monseigny, "Les Plantes du Roi," 209, n. 42.

research at the end of the century. Because comparisons based on incomplete figures can be only approximate, the conclusions offered in the rest of the chapter are tentative.

Pontchartrain spent considerably less for scientific research than did his predecessors. The impression given by the low figures gleaned from the financial records is borne out by Bignon's papers relating to supervision of the Academies. In one memorandum after another, he justified expenditure on the grounds that it was customary, necessary for the legitimate work of the Academies, and important to morale. The need to justify would have arisen only if the minister had queried requests for support. In addition, the delays of months or years which academicians endured before they were reimbursed or received their pensions also point to a more stringent financial policy during this era when the War of the League of Augsburg dominated royal spending.[5]

Overview of Research Subventions. Table 5 lists all known research subventions from 1690 through 1699, according to the fiscal year in which they were paid. Table 6 totals those data year by year according to category of expenditure, distinguishing between shared expenses and those intended exclusively for the Academy. Table 7 summarizes Pontchartrain's research subventions according to whether they benefited research in natural philosophy or in the mathematical sciences, and shows what proportion of the total research budget each individual and composite category represented.

The research subventions from 1691 through 1698, for which we have the most precise descriptions, tended to favor natural philosophy (which included chemistry, anatomy, botany, natural history in general, and mineralogy) over the mathematical sciences (which included astronomy, mechanics, and pure mathematics). The chemical laboratory appears to have been the most costly item in the Academy's research budget, constituting 38.6 percent of the total; it benefited the natural philosophers. The second major expense was for scientific instruments and models, accounting for 20 percent of the expenses; instruments and models were primarily for work in the mathematical sciences. The Observatory and "small expenses" represented 13 percent and 14.5 percent respectively of the Academy's research budget, and while the former benefited the mathematical sciences more than natural philosophy, expenditure for small expenses assisted both. The remaining categories—engravings, the *petit jardin*, and research on minerals, anatomy, and botany—together corresponded to about 14 percent of the budget and were mostly associated with natural philosophy (table 7).

Examining how the money was spent—starting with small expenses, which benefited all the Academy's work, and considering in turn expenditure on the mathematical sciences and natural philosophy—suggests that

---

[5] BN MS. Clairambault 566: 186–94, contains several memoranda from Bignon to Pontchartrain explaining and justifying expenditure for the Academies.

funding emphasized maintenance and modest, *ad hoc* reimbursements rather than the large-scale, long-term projects favored by earlier ministers.

Small Expenses (Category 8). Throughout the century, a few academicians (and sometimes also outsiders) were reimbursed for what the financial records characterized as "small expenses." Few details are provided, but these sums included equipment, supplies, repairs, transport, books, maintenance, and miscellaneous items, many intended for the Observatory. An entry for 1691 notes, for example, that Couplet incurred "small expenses . . . for the maintenance of the sites and for the exercises and experiments of the Academy."[6] In February 1696 Couplet's responsibility to pay the Academy's suppliers and seek reimbursement from the treasury was formally recognized by designating him treasurer, a title which Fontenelle later ridiculed as "fatuous and inappropriate" because "he was rather the opposite of a treasurer." Couplet "controlled no funds but made advances that were rather considerable in view of his personal wealth, and was reimbursed only with difficulty."[7]

Although Couplet was the official recipient of payments for the Academy's small expenses, other academicians were reimbursed for modest sums described vaguely in the records, and these have also been categorized as small expenses. In 1693 Chastillon, Colson, and Homberg split 892 livres. No doubt Chastillon was paid for drawings and engravings, Colson for assisting anatomical research, and Homberg for his research in the laboratory. In 1698 Fontenelle collected 609 livres 5 sous, 191 livres 2 sous of which were owed to Bourdelin for the previous year's work.

In 1699 there was a dramatic increase in the amount paid to Couplet, and in the absence of explanatory details these disbursements have been categorized as small expenses. Couplet received 12,650 livres, and Fontenelle was also reimbursed 500 livres. For that year, therefore, the Academy's small expenses came to 13,150 livres. This was more than three times as much as the Academy spent for all of its research in any one of the preceding eight years (table 6). It also exceeded the amount allotted annually for the costs of research in the eighteenth century.[8] This large sum probably included expenses incurred from 1690 to 1698 by academicians, who had to wait until the war ended to be reimbursed. But it may also represent the infusion of research support necessary to invigorate the institution immediately after the reorganization.

Small expenses accounted for 14.5 percent of the known costs of research from 1691 to 1698, amounting to 2,795 livres or an average of 349 livres annually. This is comparable to Louvois's expenditure of 2,370 livres or

---

[6] *CdB*, 3: 584.

[7] Fontenelle's eulogy of Couplet, *Histoire . . . 1722*, 128, points out that Couplet's son took the post after his father's death, but kept bad records and was accused of collecting more than he was entitled to; he in turn complained that he was losing money. See also AdS, dossier "Pierre Couplet"; Wolf, *Observatoire*, 42–44, 94–95; Cassini, *Anecdotes*, 290, 304; and Perrault, *Mémoires de ma vie*, 47.

[8] Bertrand, *Académie*, 97.

296 livres a year, but it is equivalent to only 28.5 percent of Colbert's average expenditure of 1,224 livres a year, totaling 22,040 livres. Small expenses represented 2.3 percent of Colbert's and 3.6 percent of Louvois's total expenditure for the Academy's research; excluding the cost of the Observatory from the calculations, small expenses represented 10 percent of Colbert's and 4.5 percent of Louvois's expenditure on the Academy's research. It is a sign of the decline under Pontchartrain that small expenses should represent a higher percentage of total expenditure.

The Observatory (Category 9). The Observatory was the principal seat of the mathematical sciences in the Academy. In it were housed Cassini, other astronomers, and mathematicians, as well as much of the Academy's equipment and its collection of models. Its construction was one of the principal expenses under Colbert, who used 713,704 livres, or 76.2 percent of his total direct expenditure on the Academy (excluding pensions), to build and maintain the Observatory. Louvois spent 12,335 livres on it, 18.9 percent of his total direct expenditure on the Academy (excluding pensions), principally to install the wooden tower from Marly.[9] Pontchartrain spent comparatively little on the Observatory—13 percent of his total direct expenditure from 1691 through 1698 (table 7) on the Academy (excluding pensions)—although he had to pay for work that was done in the 1680s.[10] The principal recurring expense at the Observatory during the 1690s was for the salary and uniform of the porter who lived there, rather than for the work of the astronomers.

Scientific Instruments and Models (Category 5). Most of the scientific instruments purchased for the Academy during the seventeenth century were destined for the astronomers, who needed an expensive range of equipment. But during the 1690s payments for scientific instruments represent little more than the standard retainers for maintaining mathematical instruments and pendulum clocks, and after 1693 even the retainer paid to Gosselin and Lagny disappears from the record.[11] In 1699 one of the Academy's burning mirrors was polished for 110 livres.[12] Only two items reveal purchases of new equipment. In 1696 the treasury paid Nicolas Hartsoeker 420 livres for six telescope lenses destined for the Observatory.[13] In 1698, another 1,250 livres were spent for machines, tools, repairs to the apparatus at the Observatory, and a pendulum clock. The latter was provided for

---

[9] Wolf (*Observatoire*, 15–16) calculated its cost at 713,954 livres 15 sous 11 deniers from 1667 through 1683. The difference of 250 livres 12 sous between Wolf's and my totals is explained by errors of computation in his chart (p. 15), his inclusion of one item twice, and his omission of another item, and to a difference of judgment as to whether one payment is better interpreted as pertaining to the Observatory or to the purchase of scientific instruments.

[10] Table 5, fiscal years 1694 and 1698.

[11] In some documents the latter's name is given as Tanguy.

[12] This mirror, made by La Garouste, cost the crown more than 1,300 livres in 1685 and 1686: *CdB*, 2: 592, 666, 859, 915, 1002.

[13] This item is interesting because it corrects the assumption made by Hahn (*Anatomy*, 20) that the Protestant Hartsoeker ended his association with the Academy immediately after the Revocation of the Edict of Nantes. It also supplements Wolf's history of the Observatory.

Pierre Couplet, who took it to Portugal and Brazil to observe eclipses and the satellites of Jupiter.[14] The air pump Homberg made never appeared in the treasury accounts, either because it was his own or because payment was subsumed under a heading which obscures it. Only 3,870 livres of expenditure on scientific instruments and models have been identified from 1691 to 1698, of which 3,470 livres were for instruments. This is in contrast to the nearly 58,000 livres which Colbert spent and the 24,623 livres which Louvois paid out for such apparatus. Although the picture is less gloomy than traditionally believed, nevertheless during the 1690s very little was spent to improve the Academy's equipment.

The remaining 400 livres in this category were for the Academy's collection of models. Colbert began the collection by spending at least 11,000 livres, mostly for models of agricultural and hydraulic machines, and Louvois spent another 150 livres for a model. In 1698 Pontchartrain paid Couplet for building a model of the Samaritaine pump. This sole payment suggests a modest continuation under Pontchartrain of a traditional utilitarian interest of the Academy.

The Mathematical Sciences (Categories 5, 9, 12). Under Pontchartrain expenditure on astronomy and the other mathematical sciences was greatly reduced, representing only 35 percent of the research budget of the Academy from 1691 through 1698. Yet the astronomers and mathematicians were prolific, and several publications reflect expeditions in Europe and the Far East. This apparent paradox is explained by three facts. First, many of the publications represent earlier work, such as La Hire's edition of the manuscripts left by his deceased colleagues or the reports of the Jesuit missionaries to the East.

Second, theoretical mathematics did not require expeditions, instruments, or other expensive research facilities. What the mathematicians needed was moral support, and this Pontchartrain gave by increasing their numbers in the Academy. They were also stimulated by controversy, for mathematicians in the Academy were divided into two camps, with Rolle representing Cartesian algebra and L'Hospital, Varignon, and Carré the Leibnizian calculus.[15] The result was that two mathematicians—Rolle and Varignon—who had been academicians before 1691 showed an increase in productivity from 1692,[16] and as a group the mathematicians in the Academy became more active than they had been previously.[17]

Third, some of the works which redounded to the credit of the Academy were based on expeditions funded without reference to that institution. Academicians, for example, trained the Jesuit missionaries who went to

---

[14] Unfortunately, he was shipwrecked on the return voyage and lost his instruments, journals, and specimens: *Mémoires . . . 1700*, 172–78. Pingré, *Annales*, 575.

[15] Greenberg, "Mathematical Physics," 59–60.

[16] On the rise in productivity of Rolle and Varignon, see Saunders, *Decline and Reform*, 94.

[17] According to Saunders (*Decline and Reform*, 67), mathematics and mechanics accounted for only 3 percent of the presentations at the Academy's meetings under Louvois, 10 percent under Colbert, and 14 percent under Pontchartrain.

India, Siam, and China using science as their passport.[18] Their association with the Academy began during the 1680s, when its members taught them the methods of astronomical and natural historical research and tested their instruments. Their reports began reaching the Academy during Louvois's ministry. Cassini and La Hire analyzed their data and the Jesuit Thomas Gouye edited two books, which were published under Louvois and Pontchartrain as works of the Academy.[19] Although the Jesuits remained abroad,[20] their work resulted in no new publications for the Academy during the 1690s.

Individual academicians voyaged to various locations in Europe. Whatever the initial purpose of these trips, they were often put to scientific uses. From 1694 to 1696 Cassini and his son traveled in Italy and France and made astronomical observations, and the elder Cassini took advantage of diplomatic opportunities in 1697 and 1698 to study the skies in Holland and England.[21] The two also used their journeys to record the declination of the magnetic needle.[22] Couplet's son traveled to Portugal and Brazil to make observations.

Cassini and his team of astronomers, which totaled six academicians by 1698 and outnumbered academicians in any other category, also worked in the Observatory and collected data from provincial correspondents and associate members of the Academy. They trained and sponsored the team that prepared the *Neptune françois*, and Cassini's name appeared on the title-page of that book. But above all they waited for approval to recommence the two great projects which had been suspended since the early 1680s, the extension of the meridian and the mapping of the kingdom.[23]

The Laboratory (Category 6). Chemistry was a pivotal subject in natural philosophy, one which was coming into its own as a separate discipline

---

[18] See Gallois's "Extrait du livre intitulé *Observations physiques*," 10: 130.

[19] Gouye, *Observations . . . de Siam* (1688) and *Observations . . . des Indes et de la Chine* (1692), the latter summarized by Gallois in "Extrait du livre intitulé *Observations physiques*." For a water-color portrait of the missionaries observing an "eclipse de soleil à Siam en 1688 au mois d'avril," see *Usages du royaume de Siam* (1688), BN, Cabinet des Estampes, reproduced in and discussed by Jacq-Hergoualc'h in *Phra Narai et Louis XIV*, no. 99.

[20] Père Antoine Verjus received 9,200 livres a year to support "twenty Jesuit missionaries and mathematicians" sent to the "East Indies and China for the propagation of the true faith and for the perfection of the arts and sciences": AN G^7 893, 894, 895, 897, 899, 901; 996, 9 June 1696. Given the dual purpose of their sojourn and lacking details about how the money was spent, these sums have been excluded from calculations of expenditure on behalf of the Academy. Like Hartsoeker and Cassini (see chap. 6) these missionary-scientists were also involved in clandestine diplomatic activities: see AN C^1 22: 177–78, 183–84, a memoir transmitted by père Tachard to père de La Chaise, summarized by Jacq-Hergoualc'h in *Phra Narai et Louis XIV*, no. 52.

[21] The voyage to Holland took place during negotiations for the Peace of Ryswick, and Cassini escorted the wife of the chief negotiator for the French to Holland before setting up his telescopes. See chap. 6 on the implications of his movements.

[22] *Histoire*, 2: 264; *Mémoires*, 7, 2: 461–572, containing their *Observations astronomiques faites en France et en Italie en 1694. 1695. & 1696.*, and J.-D. Cassini's *Observations astronomiques faites en Hollande, et en Angleterre, en 1697. & 1698.*

[23] Saunders, *Decline and Reform*, 79–80, 159, 164–65. Appendix B, documents VII and VIII, reveal funding of these projects in the eighteenth century.

and yet was ancillary to natural history and medicine. Furthermore, it was an important theoretical tool for natural philosophers, many of whom believed that chemical reactions provided the ultimate scientific explanation. Under all three protectors chemistry played a prominent role in the Academy, and under Pontchartrain the chemical laboratory was the most expensive single category in the Academy's research budget. The laboratory sustained research in chemistry proper and served anatomical, botanical, and other studies; during the heyday of astronomical expeditions it had also manufactured the medicines the astronomers took with them on their voyages.[24]

When Bignon began to preside over the Academy, one of the reports which informed him about its history, composition, activities, and budget also described the laboratory. It noted that the costs of chemical experiments were traditionally paid on the basis of certified *ordonnances* presented by Borelly and Bourdelin, which varied but never amounted to more than two thousand livres a year. The records of the buildings account confirm this estimate.[25] But after Borelly died in 1689 and until Homberg was appointed at the end of 1691, the laboratory at the Bibliothèque du roi stood empty, because Bourdelin worked at home, too old and infirm to conduct his accustomed level of work in the Academy's facilities.[26] As a result, the cost of chemical research fell during Louvois's ministry.

When Pontchartrain appointed Homberg, he gave him control over the idle laboratory and charged him to review Bourdelin's work, which Homberg began to do immediately, as his early reports demonstrate.[27] But financial data for Homberg's expenses in 1691 and 1692 have not been found, and the expenses known for 1693 and 1694 are so low as to suggest that they are incomplete. During the first years of his appointment, either Homberg did little laboratory research, or he worked elsewhere and did not seek reimbursement from the treasury, or his reimbursements are not identifiable in the records. From 1695 to 1698, however, the treasury reimbursed Homberg directly and quarterly. During these four years Homberg spent from eleven hundred to fifteen hundred livres annually, less than was spent from 1676 to 1683.[28] His frequent, varied, and detailed reports, not only on the previous work of the laboratory but also on phosphorus and the so-called "tree of Diana," together with his reputation for enlivening meetings, suggest that his activities were considerable and therefore that these costs are incomplete, unless he was drawing on research performed

[24] BN MS. n. a. fr. 5147: 110r.

[25] BN MS. Clairambault 566: 251v. See Stroup, *Company*, chaps. 4–5 on the laboratory.

[26] BN MS. n. a. fr. 5147: 117r; AdS, Reg., 10: 95v.

[27] Stroup, "Wilhelm Homberg." As Wolf has pointed out, there was no post formally known as the directorship of the Observatory until the eighteenth century. Likewise, not until Homberg's time do the sources refer to a directorship of the laboratory.

[28] Taking charge of the laboratory entailed accepting responsibility for its expenses. Throughout 1695 Homberg submitted *ordonnances* for small amounts, about 200 to 300 livres at a time, every three months: AN G⁷ 898. In the absence of Homberg's notebooks of expenses, it is not clear whether he hired a laboratory assistant as Bourdelin had done.

before he entered the Academy. Another problem is that no expenses are reported for either Charas or Boulduc, both of whom conducted chemical research as academicians during the 1690s. They too relied on other laboratories, for given Charas's age he, like Bourdelin, preferred to work at home,[29] and Boulduc had access to facilities at the Jardin royal, where he taught the course in chemistry. Nevertheless, both academicians would have been entitled, under normal conditions, to reimbursements.

The evidence about Bourdelin's expenses supports the supposition that figures for the laboratory are incomplete. Bourdelin's notebook of expenses from 1666 until his death in 1699 has survived, and it supplements the buildings accounts and royal treasury records in several ways. For example, Bourdelin was occasionally reimbursed by intermediaries—that is, by Gruyn in 1695, by Bignon's secretary in 1696, and by Fontenelle in 1699—but only one such payment has been identified in the records of the treasury. Without Bourdelin's invaluable notebook of expenses, estimates of the costs of maintaining chemical research would be too small. Unfortunately no such notebook has been found for Homberg's regime at the laboratory. Judging from levels of previous expenditure, from Homberg's record of activity, and from instances where other documents supplement the treasury accounts, the known expenses of the laboratory are incomplete. Nevertheless, the 7,414 livres paid from 1691 through 1698 represent 38.6 percent of the Academy's research budget and constitute the largest known commitment of resources to the Academy's research in these years.

Natural Philosophy (Categories 4, 6, 7, 10, 11, 13, 14). The Academy published more astronomical and mathematical than natural philosophical works during the 1690s, but Pontchartrain did not follow Colbert's example by spending more on astronomy than on any other aspect of the Academy's work. Like Louvois, he was more likely to authorize funding for natural philosophical research. Natural philosophy absorbed 50.5 percent of Pontchartrain's known direct disbursement on academic research from 1691 to 1698, and at their meetings academicians presented more reports on natural philosophy than on astronomy and mathematics during this period.[30]

The study of natural philosophy required four principal forms of support which resulted in recurring expenses: the *petit jardin*, the chemical laboratory, corpses for dissections, and engravings. The first three were essential for research, while engravings were necessary for publication. There were also small, specific outlays to academicians for botany, anatomy, and mineralogy.

The crown spent roughly 100 livres a year on the *petit jardin*, which the Academy's botanists used. Part of the Jardin royal, it was controlled by Jean Marchant until 1694, when Guy Crescent Fagon wrested it from him.

---

[29] *Histoire*, 2: 152 (1692).

[30] Saunders (*Decline and Reform*, 172) counts the number of scientific presentations in various subjects, but his calculations cover the period 1692 through 1699.

Since Fagon's antipathy toward Marchant did not extend to the Academy, the *petit jardin* may have continued to serve the Academy even after 1694. Under Louvois it had cost only 40 livres yearly, but in the 1690s the cost of cleaning the amphitheater was added to the sum for maintaining this small garden.[31] In addition to these recurring costs, Tournefort was reimbursed 118 livres in 1697, and two years later Marchant got 364 livres, no doubt also for botanical research.

Like botany, anatomy linked the Academy to the Jardin royal. Du Verney worked there and charged to the budget of the Jardin some of his work on behalf of the Academy. His anatomical research for the Academy cost at least another 556 livres.

The botanists and the anatomists depended on the laboratory. But whereas the former needed it to analyze plants and their constituents, Du Verney and Méry required it to distill brandy, which they used during dissections to clean putrid matter from the viscera of bodies, to wash their hands, and to drink for refreshment.[32]

The anatomists needed bodies to dissect, and having completed their dissections they wanted the skeletons mounted. Colson had met both of these needs in the past and probably continued to do so in the 1690s, thus earning part of the 892 livres paid in 1693. During the 1690s Du Verney and Méry dissected a sea tortoise, a dog, a young man struck by lightning, an infant, the paws of a lion, an ostrich, a Spanish fox, at least two pelicans, five porcupines, a lizard, several vipers, and other human and animal subjects. Du Verney's 1692 article on the routes of bile and the pancreatic juices drew on his dissections of the five porcupines and two ostriches. Méry's articles of 1692 and 1693 on the circulation of the blood in the fetus and the problem of respiration cited his dissections of a sea tortoise, and he also published an article on the skin of the pelican. Many of the Academy's corpses had traditionally come from the royal menagerie, giving rise to inconsequential costs of transportation from Versailles to the Bibliothèque du roi and on to Du Verney's lodging at the Jardin royal or to Méry's at the Invalides. Unfortunately, these costs, which amounted to about 60 livres under Louvois, are not known for the 1690s.[33]

---

[31] Laissus and Monseigny ("*Les Plantes du Roi*") connect Marchant's loss of the *petit jardin* to the publication of Tournefort's *Élemens*.

[32] BN MS. n. a. fr. 5147: 12r.

[33] *Histoire*, 2: 117–339; BN Archives de l'ancien régime 2: 1r and Clairambault 814: 643–45, show that in 1692, the Bibliothèque du roi paid for the 1691 transport of an ostrich and a civet from Versailles to the Library and then to the Jardin royal. For porterage of dead animals from Versailles to academicians during the 1680s, see Archives de l'ancien régime, 1: pp. 45, 68, 77; fols. 49v, 66v, 67r, 73v, 74v, 75v, 77r, 78r, 80r, 81r, 83r–v, 85r, 87r, 90v. Aside from an ostrich, these animals were not the subjects of Méry's and Du Verney's papers of the 1690s. Thus, Pontchartrain was publishing articles based on anatomical work done during his own protectorship rather than Louvois's. See *Mémoires*, 10: 26–27, 65–67, 271–75, 324–25, 386–97, 433–38, for Méry's and Du Verney's articles. It was Auzout who in March 1667 suggested that when birds and animals died at Versailles they be transported to the Academy for dissection: AdS, Reg., 1: 204.

Contemporaries believed the crown spent 12,000 livres for the plates to Tournefort's *Élemens;* this was plausible, for illustrations were costly and Colbert and Louvois had spent about twice as much on engravings for the Academy's natural history of plants. An expenditure of 12,000 livres would dwarf known expenditure for other research and would have reduced the funds available for the Academy's other projects. Since the *comptes du Trésor royal* contain no reference to the cost of Tournefort's book, however, that sum has been excluded from the present calculations. Otherwise, the expenses of illustrating the Academy's publications about plants and animals were largely a thing of the past. Pontchartrain continued the practice of keeping the engraver Chastillon on retainer for the Academy. In addition Joubert received 547 livres for drawings and engravings of plants, and Chastillon got 319 livres for depicting animals.

If the research subventions for studies of plants and animals seem conventional, the money given to the now obscure Morin de Toulon for his experiments on minerals represented a new departure for the Academy. Morin's official title after 1699 was "botanist," but references to his work in the 1690s show only research on minerals and soils. He first appeared in the histories of the Academy in 1693 with his paper on a vein of soft iron. That year he also presented the plans for a study of minerals. The following year, he brought to one meeting a bone which he had found in the Montmartre lime quarry and which Méry identified as the rib of a very large tortoise. Morin also analyzed the European imitations of Chinese porcelain and discussed the composition of the earths required to obtain the best product. His report in 1694, on the "azur de cendres bleuës" he found during his exploration of Mont d'Usson in the Auvergne in 1688, mentioned medical uses and closed with a brief proposal for more work on minerals. His research cost six hundred livres from 1694 through 1697.[34]

Under Pontchartrain research in natural philosophy was a complex blend of old and new. The natural history projects familiar under Colbert and Louvois—the illustrated guide to plants and comparative anatomy of animals—had been crippled by uncertainties about the research or by loss of personnel. Dodart, Marchant, and Bourdelin continued to study plants in the old way, and Homberg at first reacted favorably to their research, which he had been ordered to review. But Homberg became more skeptical about the chemical analysis of plants which Dodart and Bourdelin championed, and he joined the ranks of those who pursued individual interests rather than the traditional team projects. Likewise, without the tutelage of Claude

---

[34] *Histoire*, 2: 183, 205, 208. Mont d'Usson is probably the escarpment in the Puy-de-Dôme where the village called Usson is located. On the geology of the area, see Girault de Saint Fargeau, *Dictionnaire géographique*, 3: 711; Ministère de l'Industrie, "Carte des richesses minérales de la France" and Ministère de l'Industrie et du Commerce, "Carte géologique de la France." See AdS, dossier "Morin de Toulon," for a memoir dated 23 January 1694 on *azur.* "Cendres bleues" were found in soft stone in areas having copper deposits: Diderot, *L'Encyclopédie*, 2: 284, on "cendres bleues" and 12: 579, on "pierre d'azur."

Perrault, who had been the driving force behind the natural history of
animals, the anatomists concentrated on other studies, worked independently from one another, and did not even see Perrault's final volume to
completion, even though it was partially printed and some new illustrations
were prepared for it. Du Hamel had probably taken responsibility for this
work after Perrault's death, for in 1690 he was reimbursed for expenses
incurred from 1688 (when Perrault died) through 1690 on behalf of the
continuation of the project, and when he resigned as permanent secretary,
he took the title "anatomist." Work on the history of the Academy may
have distracted Du Hamel from seeing the final volume of Perrault's natural
history into print before 1700.[35] Morin's studies of minerals are typical of
the new trend, for he designed and researched his project independently,
although he relied on the Academy to referee his papers and on the treasury
to fund his research in a modest way.

Publications. One of the most invigorating of Pontchartrain's directives
to the Academy was that it produce more books and articles. Over the
years academicians had unsuccessfully requested permission to publish
their research findings, with the result that by 1691 there was a backlog
of unpublished manuscripts and research notes. Pontchartrain not only
granted academicians permission to print works dating from earlier years,
but also commanded them to publish monthly papers. As a result, during
the 1690s the Academy published numerous books and articles, using both
the Imprimerie royale and private Parisian printers.

Under Colbert and Louvois academicians had written more than they
were permitted to publish, but under Pontchartrain the tables were turned,
for he ordered them to publish even more than they could write. After
1693, academicians were unable to sustain the monthly articles. L'Hospital
explained their failure to Martin Lister as due to inadequate correspondence.[36] To an English savant, whose principal point of reference would
have been the *Philosophical Transactions,* this was a telling excuse, for the
English journal did not rely solely on the writings of Fellows of the Royal
Society. L'Hospital's explanation, however, was disingenuous, for it told
only part of the story. While the Academy was indeed smaller than the
Royal Society, and its correspondence was insufficient to fill a monthly
journal, it was beset by other problems, such as delayed pensions and a
reduced research budget, which undermined the morale of members and
contributed to the Academy's inability to publish articles every month.

What little is known about costs suggests that the crown spent more in
the 1690s on the publication than on the research of academicians. In the
last decade of the century, the Imprimerie royale worked on more than a
dozen volumes related to the Academy, including Tournefort's *Élemens*
and *Histoire des plantes,* La Hire's edition of manuscripts left by deceased

---

[35] Saunders (*Decline and Reform,* 66–77, 164) blames financial hardship for Pontchartrain's
failure to publish this work.

[36] Lister, *Journey,* 80, 97.

academicians, L'Hospital's treatise on the infinitely small, the reports of the Jesuits, several of Cassini's works, the eighty-five articles published in 1692 and 1693, the natural history of animals, and the *Neptune françois*.[37] These publications bore little relation to the budget for research during the 1690s, for most stemmed from work done either outside the institutional structure of the Academy or during the 1670s and 1680s. Tournefort's *Élemens* fell into the former category, La Hire's edition of his colleagues' manuscripts into the latter. Pontchartrain earned the gratitude of academicians for redeeming their earlier work by ensuring its publication, but he was more successful in publishing previous than new research.

Conclusion. Three trends are apparent from the Academy's research budget during the 1690s: the crown reduced its financial support for research, much expenditure was for maintenance of buildings and equipment rather than for research itself, and the costs of research bore little relation to what was published in this period.

Seen from the perspective of its halcyon days under Colbert, the research program of the Academy during the 1690s represents a decline. The policy of austerity forced on Pontchartrain undeniably injured the institution in many respects. Given the weakened state of the Academy when Pontchartrain became its protector in 1691, the decline of the institution after 1695, and the poverty of the royal treasury during the 1690s, the question arises why the crown preserved the Academy at all. Two features of the Academy, its usefulness and its relative cheapness, helped guarantee its survival.

---

[37] Bernard, *Histoire de l'Imprimerie royale*, 138–54; Saunders, *Decline and Reform*, 212–20, 276–78.

PLATE 2. Abbé Jean-Paul Bignon, President of the Académie Royale des Sciences. Portrait by Lucretia Cath. de la Roue, engraved by Edelinck, 1700.

VI. UTILITY AS A GOAL

Like the academicians, Pontchartrain and Bignon respected knowledge for its own sake, and they tried to protect and direct the republic of letters, including the Academy of Sciences.[1] They guaranteed the survival of the Academy by continuing to fund it, and some of their measures at least temporarily improved institutional morale. They also repaired the Academy's image by increasing membership and printing books and articles by academicians, hence creating an illusion of institutional vigor. But in reality funding was reduced, many academicians did not attend meetings, and several publications represented old research. Moreover, the fact that the Imprimerie royale published much of the work of academicians suggests that the Academy's most important audience in this crucial decade was the king, for the publications of the royal printing house were normally presentation copies, not for sale in bookshops in the ordinary way. To survive, the Academy had to enhance the king's *gloire*, and elegantly produced books filled with discoveries obtained under royal auspices did just that.

The Academy and its protectors had still another strategy, one which would appeal to a patron who had no head for scientific theories, but who wanted better maps, better health, and better military equipment. By showing that their work could benefit king and kingdom in such practical ways, academicians could demonstrate to their royal patron the value of the Academy. This strategy was consistent with their own notions about natural philosophy and mathematics, for in common with other savants of the age, academicians believed that theoretical research would have practical consequences.

The minutes and publications of the 1690s show that the Academy pursued applied science in many forms. Mathematicians and astronomers focused their utilitarian attentions on navigation and cartography. In his audiences with Louis XIV, Cassini repeatedly assured the king that astronomical work would improve these spheres of knowledge so important to a ruler. Data collected at the Observatory, by Jesuits in the Far East, or by Cassini's provincial and foreign correspondents were all used to correct the longitude of major cities, as the map on the Observatory floor witnessed.[2]

---

[1] Saunders, *Decline and Reform*, 138–48, 154–55, 160.

[2] Cassini, *Anecdotes*, 291, on the astronomer's assurances to the king; on reforming the map of the world, see *Histoire*, 2: 265. The ephemeral map inked onto the floor was preserved in an engraving by J. B. Nolin in 1696: Pelletier, "Les globes de Louis XIV," fig. 4.

Botanists and chemists sought to improve medicine, and the minutes and official histories of the Academy are filled with discussions of remedies. In 1694, for example, Charas presented papers on medicaments, especially opium, and in the following year Homberg proposed a way of improving cataract operations.[3]

Manufacturing, horology, and calendrical reform also found a place in the Academy's interests. Homberg studied dyes, and Morin de Toulon's report on porcelain is a chemist-mineralogist's study of a process in which European royalty invested heavily during the seventeenth and eighteenth centuries. Cassini reviewed a proposal for reforming the calendar, and Varignon took an interest in improving clocks.[4]

The Academy also had an educational function, realized not only through its publications but also by its room of machines at the Observatory. This was open to the public for the study of natural philosophical apparatus such as the burning mirror; models of industrial, military, and agricultural machines; astronomical instruments; and the map of the moon prepared by Patigny under Cassini's direction.[5]

The Academy had begun under Colbert an ambitious project of mapping the entire kingdom, but funding lapsed in the 1680s for the extension of the meridian and other surveying necessary for the task. Although Pontchartrain did not recommit the treasury to this project until after the reorganization of 1699,[6] he did support a less well known cartographical project in which the Academy played a role. This was the *Neptune françois* published in 1693, a collection of maps which surveyed the European coasts from Trondheim in Norway to the straits of Gibraltar.[7] Cassini trained several of the men who worked on it, including Chazelles, Sauveur, de La Voye, de Gennes, and de Pène, who were associated with the Academy in other projects. La Voye had actually been a student member briefly in the 1660s, and Chazelles and Sauveur became academicians in 1695 and 1696.

The treasury records shed light on the *Neptune françois*. A petition from Sauveur and de Pène to the king survives among the treasury's papers for the 1690s. In it the two explain that their ambitious project of mapping the seas around France, including the Mediterranean, was expensive. They had spent fifteen thousand livres on, among other things, "reduction, copper, engraving, paper, and printing," of which they expected to recoup only five hundred livres through the sale of the printed maps. For that reason the engineer and the mathematician applied to the king for a sub-

---

[3] *Histoire*, 2: 117–18, 211–12, 245–46; Stroup, *Company*, chap. 13.
[4] *Histoire*, 2: 236–38, 277, 300–31, 334, 340.
[5] Wolf, *Observatoire*, 96–97, 129, 168–70. According to the *CdB*, that map cost 7,185 livres in payments to Patigny, who worked on it from 1672 to 1679.
[6] Appendix B, documents VII and VIII. For efforts to obtain cooperation from the king of Spain, see *Correspondance administrative*, 4: 623–24.
[7] The introduction included Sauveur's *Explication des echelles*.

vention. The result was that the *Neptune françois* was published simultaneously by Jaillot and the Imprimerie royale.[8] The Academy then planned a second volume to map the Mediterranean. Thus, Chazelles voyaged "par ordre du Roi" to calculate the latitude and longitude of major cities along that sea and sent his data to the Academy. But like so many other projects of the Academy, the continuation of the *Neptune françois* never appeared in print.[9]

Under Pontchartrain academicians continued to find practical occupations outside the institution. Homberg was one of four persons (the others being a carpenter, an ironworker, and a painter) who shared 398 livres 10 sous for "pieces of work they supplied to the king."[10] Many members of the Academy taught mathematics, which was necessary for fortification and cartography: La Hire and Sauveur lectured on mathematics at the Collège royal; Sauveur tutored the son of the duke of Orléans and the grandsons of Louis XIV, the dukes of Anjou and Berry; and C.-A. Couplet, and later his son P. Couplet, taught mathematics to the pages of the Grande écurie.[11] Sauveur also acted as purchasing agent for his charges, ordering instruments and buying geometry treatises.[12]

The Academy's interest in technology is epitomized by its studies of machinery, which date back to Colbert's protectorship. Throughout the century academicians and their paid associates[13] designed and tested apparatus, and outsiders also submitted inventions to the group, hoping for its approval.[14] When Pontchartrain took control of the Academy, the Observatory's collection of machines boasted models of ancient military equipment, including catapults, and all sorts of machines used in the arts and crafts, such as one for making fabrics and another for unwinding

---

[8] *Neptune françois*; AN G⁷ 992; Bernard, *Histoire de l'Imprimerie royale*, 15, and Bourgeois and André, *Sources de l'histoire de France*, 1: no. 198. I am grateful to Anthony Turner for drawing my attention to Sauveur's book and biography. On de Pène, see Colbert, *Lettres*, 5: 176–77, n. 1.

[9] *Histoire*, 2: 222–24; Saunders, *Decline and Reform*, 203; *NBU*, 10: 169; *DBF*, 8: 955–56; on the increasing importance of the Mediterranean in French naval strategy after 1690, see Symcox, *Crisis*, 108–17.

[10] AN G⁷ 894, May 1692, PC.

[11] On the pages of the Grande écurie, see *État de la France* (1699), 547–53, and Willems and Conan, *Liste alphabétique*, 17–19.

[12] The books cost 590 livres: AN G⁷ 901, Jan. 1698, PC, and 902, Sept. 1698, PC. Sauveur probably ordered the instruments Chapotot made for the dukes; Chapotot was paid 462 livres 15 sous for these: AN G⁷ 901, Mar. and May 1697, PC.

[13] From the 1680s Dalesme received a yearly pension of 600 livres for inventing machines. He became a member of the Academy in 1699. Although his first name is given as André in the *CdB*, he may be the Jacques Dalesme known to the duke of Roannez and his family; see Mesnard, *Pascal et les Roannez*, 925–26. Leibniz knew Dalesme as the "inventor of the pneumatic tubes for carrying forces a great distance": Leibniz, *Philosophical Papers and Letters*, 2: 771. See also Blegny, *Livre commode*, 1: 76; *Journal des sçavans* (1 Apr. 1686); Nicéron, *Hommes illustres*, 10: 180. Dalesme invented machines for the mint, AN G⁷ 997 (1699), and was awarded an exceptional pension of 1,500 livres in 1692: AN G⁷ 894, June 1692, GC.

[14] For the argument that by examining new inventions the Academy served the public interest, see *Histoire*, 1: 14.

bobbins. James II of England visited the Observatory in 1690 and looked at several models of capstans, apparatus for lifting and balancing weights and raising water, and a portable bridge which Couplet had invented.[15]

Some of the inventions the Academy considered in the 1690s had a decidedly military purpose, as when Amontons presented a design for flexible, cheap, and lightweight pontoons, or Dalesme demonstrated the recoil of a cannon. A visitor to the Observatory in 1696 came away impressed with the usefulness of the *salle des machines* to military engineers and teachers of fortification and navigation. Even Varignon's paper of 1696 about why bodies of drowning victims rise to the surface drew on a battlefield analogy. Several members had military or naval experience. The marquis de L'Hospital had been trained in the arts of warfare as a matter of course. The admission of Chazelles to the Academy as an associate astronomer in 1695 followed his exploits in the naval campaigns of 1687, 1688, and 1690; as professor of hydrography in Marseille, based at the arsenal of the galleys, he wanted to test his theories. Likewise, when Sauveur decided to write a book about fortifications, he first sought practical experience at the siege of Mons. These academicians were no armchair strategists but rather active campaigners.[16] The king's wars affected the Academy not only by reducing its funding, but also by drawing the attention of members to military technology and tactics.

The Academy also provided cover for clandestine activities on behalf of the crown. Thus the Dutch instrument-maker Hartsoeker not only supplied telescope lenses and presented theoretical papers to the Academy, but also assisted French diplomatic initiatives for peace with Holland. He traveled between the two countries in 1692 and 1693, gathering information for the French about Dutch sentiments and identifying influential republicans favoring peace.[17] His return to Holland in 1696 did not end his association with the Academy, for Cassini visited him there the following year, after conducting Mme de Harlay, the wife of the chief French negotiator in the talks leading to the Peace of Ryswick, to her husband.[18] In view of Hartsoeker's clandestine activities and the identity of Cassini's traveling companion, it is unlikely that Cassini visited Hartsoeker purely for the sake of observing the skies from the latter's Rotterdam garden. Their scientific interests and previous scholarly association gave Cassini and Hartsoeker the perfect pretext for meeting if their services were required during the negotiations for peace.[19]

---

[15] Wolf, *Observatoire*, 96–97, 129.

[16] *Histoire*, 2: 157, 221, 277; see also Dalesme's design for a turning bridge, 262; Wolf, *Observatoire*, 96–97; *DBF*, 7: 955–56; *NBU*, 10: 169, 43: 379–80.

[17] André and Bourgeois, *Recueil . . . Hollande*, 1: 410–11.

[18] *Histoire*, 2: 330; André and Bourgeois, *Recueil . . . Hollande*, 1: 443ff. In 1699 Hartsoeker became a foreign associate of the Academy.

[19] The agent who used Hartsoeker's memoranda to assist his spying in Holland in 1692 and 1693 was the artist and writer Roger de Piles, whose cover was a search for works of art: Mirot, *Roger de Piles*, chap. 3; Teyssèdre, *Roger de Piles*, 397–99.

The Academy served the king's civil needs as well. A principal concern since the 1670s was the supply of water to the palace at Versailles. Academicians had surveyed for aqueducts under Colbert and Louvois, and from this work the savants developed books on surveyors' levels and hydrography. In the 1690s the Academy continued to study water control, for instance by reviewing Cassini's recommendations to the pope for governing the Ferrarese waters. Sédileau calculated the capacity of reservoirs and sometimes framed his papers in the context of the needs of Versailles. Even the yearly measurements of rainfall at the Observatory were justified and perhaps motivated by concern for maintaining water levels at Versailles.[20]

The Academy and its three ministerial protectors were serious about utilitarian goals, but a bald summary of accomplishments in this category perhaps overemphasizes the extent and nature of these investigations during the 1690s. By comparison with previous years, the Academy's practical research was smaller in scope and piecemeal. Under Pontchartrain it lacked both the emphasis on theoretical underpinnings that had characterized the Academy's technological research under Colbert,[21] and the commitment to massive projects of public works which both Colbert and Louvois had sponsored. Furthermore, the Academy's utilitarian activities cost less under Pontchartrain. Whereas Colbert spent 51,500 livres, or 5.5 percent of the research budget (actually 23 percent, if the Observatory is excluded from the calculations), and Louvois spent 6,143 livres, or 9.4 percent of the research budget, Pontchartrain spent only 719 livres, or 3.7 percent of the research budget from 1691 through 1698, on practical and technological projects.[22] Finally, the number of inventions submitted to the Academy for approval also declined.[23]

But the Academy was not the sole representative of the crown's practical expectations from science, for Pontchartrain and Bignon supplemented the Academy's technological functions by forming a sister-society. This was the Compagnie des arts et métiers, which also figures in the treasury payments of the 1690s. The three principal members of the Compagnie—père Sébastien Truchet, Gilles Filleau Des Billettes, and Jacques Jaugeon—enjoyed close relations with the Academy of Sciences and joined it on its reorganization in 1699, and the Compagnie's minutes were inserted in the

---

[20] *Histoire*, 2: 42, 133–35, 164–67; *Mémoires*, 10: 29–36. Sédileau's notebook and papers survive in AdS, Carton 1680–99. Louvois instructed the Academy to study rainfall in connection with the water supply of Versailles: Wolf, *Observatoire*, 110. Earlier Pierre Perrault had sent his records of rain- and snowfall in Paris from 1667 to 1668 to the Academy: AdS, Reg., 4: 123v–24v (4 Aug. 1668).

[21] Books such as La Hire's *Traité de Mécanique* and *Traité des épicycloïdes*, which appeared in 1694 and 1696, had actually been written before the 1690s.

[22] Table 5, fiscal year 1696, category 12, and fiscal year 1698, category 5.

[23] Saunders (*Decline and Reform*, 198) compared eight-year periods under the three protectors and found that from 1676 to 1683, under Colbert, the Academy reviewed forty-four inventions, as compared with thirty-three under Louvois and fifteen under Pontchartrain. Six of the fifteen were submitted in 1699.

Academy's own minutes to emphasize the connection between Compagnie and Academy.[24] Until 1699 the Compagnie formed a separate, allied group with strictly technological concerns which met on Mondays (later Tuesdays) at Bignon's *hôtel*.[25]

The mission of the Compagnie was a source of dispute. Des Billettes envisaged a group which would take over some of the functions of the Academy, such as reviewing inventions, either as a permanent subdivision of the Academy or as a separate entity.[26] He and his colleagues hoped also to compile an encyclopedia of arts and crafts, and the three interviewed artisans, sketched tools, workshops, and mills, and read treatises on technology. But their energies were from the start focused on one particular craft, printing, because of the inclusion of Anisson, director of the Imprimerie royale, and Philippe Grandjean, a type-founder, in their Compagnie. Anisson, Bignon, Pontchartrain, and the king wanted a new typeface for the royal printing house, something distinctive that was scientifically designed for maximum legibility and beauty.[27]

Disappointed at this narrowing of their charge, Truchet, Des Billettes, and Jaugeon rationalized it on the grounds that printing served all the other arts,[28] and they set out to pursue the subject exhaustively by also examining the ancillary arts of binding, goldworking, tanning, papermaking, typecasting, and typesetting.[29] Nevertheless, at least one dissenter queried why printing should be given priority, since it was a late invention that stood as no more than "a valet . . . [or] a messenger to his master."[30] Tensions between the savants and the royal printer grew, for the Compagnie des arts et métiers did not wish to be a mere consultant to the Imprimerie royale, while Anisson reasonably feared that its research would become so diffuse as to undermine work on his typeface.[31] But once they had de-

---

[24] AdS, Reg., 13: 154r–60r. Des Billettes thought the Compagnie ought to be a subsection of the Academy, and he and his colleagues thought of themselves as having been academicians even before their 1699 appointments. AdS, dossier "Jacques Jaugeon"; AN M 849, no. 8, "Disposition des planches des arts et metiers . . . gravées par . . . Simonneau."

[25] Saunders, *Decline and Reform*, 173, 196–99, on the relationship with the Academy; Salomon-Bayet, "Préambule théorique," 236, on days and place of meetings. Leibniz wrote of the relationship between the two societies, "il sera bon qu'il y ai de l'intelligence entre la soeur ainée et la cadette": *Oeuvres*, 2: 365.

[26] Saunders, *Decline and Reform*, 197; Salomon-Bayet, "Préambule théorique."

[27] Jammes, "Innovation" and *Réforme*; Steinberg, *Five Hundred Years of Printing*, 168–69.

[28] Fontenelle later conveyed this official justification to Leibniz via private correspondence (Leibniz, *Lettres*, 201), and to the public via the history of the Academy (Jammes, *Réforme*, 6–7).

[29] Disagreement over the mission of the Compagnie des arts et métiers is clear in the manuscript record. An eighteenth-century history of the Imprimerie royale by one of its directors presents Anisson's view: AN M 802, no. 1: 5v–6v. For the other side, see a note preserved among Truchet's papers, which complains that members of the Compagnie have been prevented from realizing their principal project, an encyclopedia of arts and crafts: AN M 849, no. 8. Materials collected for the encyclopedia may be seen in AdS, dossier "Truchet," and in his papers at the Archives Nationales in series M.

[30] AN M 849, no. 8; this fragment of a letter is inside a list of Simonneau's engravings for the Compagnie.

[31] A glance at the plates Louis Simonneau engraved for the Compagnie or at the notes of the three savants bears out Anisson's worry, for they addressed every aspect of the printer's art and ancillary crafts, and investigated several other arts and crafts as well: Simonneau,

signed the typeface—the famous *romain du roi*—the savants' own appre-
hensions were confirmed, for although their treatise on printing was finished
by 1704, it was never published. Having served Louis's *gloire* with the
typeface, the savants were incorporated into the Academy where they again
took up their work on the encyclopedia of arts and crafts.

Judging from such records as have been identified, the Compagnie was
costly. Truchet, Des Billettes, and Jaugeon all received pensions of one
thousand livres a year for their "application to the description and perfection of the arts."[32] For the six years from 1693 through 1698, therefore,
their pensions cost the crown eighteen thousand livres. In addition there
were payments to Simonneau for plates,[33] to Grandjean for matrices and
other expenses,[34] and to artisans for information.[35] Without complete figures
only imprecise comparisons are possible, but the crown certainly spent
more on the Compagnie than on the research of the Academy during the
same years. Thus the Academy's decreased expenditures on utilitarian activities under Pontchartrain do not signify his preference for pure over
applied science, but rather result from his experiment with a separate technological and consultative institution, which ultimately was incorporated
into the Academy, where some savants thought it belonged in the first
place. In 1699, with Des Billettes, Jaugeon, and Truchet continuing as academicians their work on the encyclopedia of arts and crafts, the crown
spent another 3,097 livres on the Academy's practical projects, in this case,
for Simonneau's plates. Thus from 1691 through 1699 the Academy's
technological interests accounted for 10.3 percent of the research budget.

Despite their different missions and separate identities, the Academy
and the Compagnie were connected even before 1699 by friendships and
shared interests among their personnel.[36] These connections show that
utilitarian interests not only were an end in themselves but also had become
fashionable in the late seventeenth century. Amateurs collected machines
and intricate toys and created a market for books on mechanics. Thus, the

---

Recueil d'estampes. See also AN M 849, no. 8; AdS, dossier "Des Billettes." AN M 851 contains
more papers which resemble those in Des Billettes's dossier, although they are catalogued as
Truchet's. On Simonneau's plates as precursors of those in the *Encyclopédie,* see Huard, "Les
planches," and Proust, *Diderot et l'Encyclopédie,* 177–78.

[32] AN G⁷ 897, 4 July 1696, GC (for 1693); 898, Nov. 1696, GC (for 1694); G⁷ 899, July,
Oct. 1697, May 1698, GC; G⁷ 901, Jan., Dec. 1698, GC; G⁷ 902, July 1699, Feb. 1700, GC.
When he joined the Academy, Truchet's pension could not be included in the *estat* of the
Academy because of the new rules, but it was continued separately: AN G⁷ 986–87, 27 Apr.
1700. See table 1, fiscal years 1700–03, table 2, and appendix C.

[33] On Simonneau, see Bénézit, *Dictionnaire critique,* 9: 618; Jammes, "Innovation," 136;
Simonneau, *Recueil d'estampes;* AdS, dossier "Des Billettes"; Salomon-Bayet, "Préambule
théorique," 234, n. 4, and 236–37; and AN G⁷ 898, Oct. 1695, PC; 899, July 1696, PC; 901,
May, Aug., Oct. 1697, Jan. 1698, PC; 902, Dec. 1698, PC. This is not a complete list.

[34] Bernard, *Histoire de l'Imprimerie royale,* 78–83, refers with little detail to the expenses of
Grandjean on behalf of the Compagnie; these principally benefited the Imprimerie. For some
of these, see AN G⁷ 898, Oct. 1695, PC; 901, Oct. 1697, Jan. 1698, PC.

[35] AdS, dossier "Des Billettes," and Salomon-Bayet, "Préambule théorique," 234, n. 4, and
236–37.

[36] AdS, Reg., 11: 119v–20v; *Histoire,* 1: 432–34; Blegny, *Livre commode,* 1: 165, n. 2; Leibniz,
Oeuvres, 1: 279–80, 288, 298; 2: 364–65; Leibniz, *Philosophical Papers and Letters,* 2: 770–76;
Malebranche, *Oeuvres,* 18–19: 58, 144–45, 158, 160, 162, 582; 20: 152, 194.

editor of a dictionary had paid artisans in the 1680s to explain their technical vocabulary,[37] just as members of the Compagnie were to pay them in the 1690s to reveal their methods.[38] Knowledge of the arts and crafts and an understanding of complicated apparatus amused gentlemen and attracted savants,[39] and royal funding reflected not only the utilitarian but also the fashionable impulse.

Academicians and members of the Compagnie shared ties of friendship, knowledge, and patronage. Since patronage was contingent on practical results, to advertise specific achievements helped ensure the survival of the two institutions. One result was that the fates of the Academy and the Compagnie were tied to the Imprimerie royale, the linchpin in the ministerial policy of publicizing scholarly activities. Although the savants did not entirely accept the primacy of printing, Pontchartrain and Bignon emphasized it in justifying, in very different ways, the existence of the Academy and the Compagnie.

A second result was that the members of these learned societies shared the frustration of seeing their scholarship channeled into a narrower role than they had expected. Savants appreciated knowledge primarily for its own sake, although they also anticipated practical benefits. But the king had more liking for concrete results than for theory, and saw utility primarily in terms of royal *gloire*. Thus the crown tended to treat the savants as technical consultants to itself or to the Imprimerie royale, whereas the savants saw themselves as benefiting knowledge, medicine, technology, and cartography in a broader sense.

Seventeenth-century scientists were savants with useful knowledge. This utilitarian potential helped justify subventions to the Academy as well as to its sister-society, even in the troubled 1690s. But the incentives were insufficient to ensure that the Academy and the Compagnie receive support at Colbertian levels. The crown supported scholarship, whether for its own sake or for more utilitarian motives, partly because it could do so at relatively little cost. It is the irony of Pontchartrain's protectorship that inadequate funding undermined the morale of the royal scholarly societies he governed, but that part of his success in preserving them was due to their relatively small budgets at a time of general financial exigency.

---

[37] Furetière, *Recueil des factums*, 2: 233, on his interviews with artisans.

[38] The difficulty of obtaining information from artisans impeded the encyclopedia of arts and crafts, according to Fontenelle: Leibniz, *Lettres*, 201.

[39] Many "persons of quality" inspected the Academy's room of machines at the Observatory, and Peter the Great visited Truchet's monastery to see his *cabinet des machines*. Truchet's own biography reveals how widespread this interest was in the late seventeenth century, for his own interest in the subject awakened as a result of a youthful visit to a famous *cabinet des machines* owned by a gentleman, and he was later to design a *galerie de mécanique* for the director of *postes: Histoire . . . 1729*, 93–101, esp. 98; Lery, "Truchet membre," esp. 18, and "Truchet en Auvergne."

VII. PONTCHARTRAIN'S MIXED RECORD AS PROTECTOR

The French crown established the Academy of Sciences as part of a program for controlling the artistic and intellectual life of the kingdom so as to benefit the monarch. Louis and his advisers wanted to improve knowledge, but they also wanted the king's name associated with any discoveries. They expected academicians to make practical advances, and they came to rely on the Academy as a consultant to the crown. When Pontchartrain became protector of the Academy in 1691, he could have allowed the institution, which was already seriously weakened, to succumb. Instead, he reinvigorated it as much as possible given the state of royal finances. Thanks to Pontchartrain's remedies, and in spite of limited funding, the Academy remained a useful instrument of state and an honorable association for its members.

Royal funding of the Academy during the 1690s is best understood through comparison of Pontchartrain's regime with Colbert's and Louvois's and of the Academy's budget with that of other Parisian institutions. These comparisons show that while the Academy was generously funded in the 1660s and 1670s, from the 1680s its finances began to decline and during the 1690s they had become modest not only in absolute terms but also relative to the finances of another royal learned institution. But first it is important to get an overview of the Academy's expenses throughout the seventeenth century.

The seventeenth-century Academy cost its royal benefactor 2,139,269 livres in direct expenditure (table 8). This can be broken down into three general categories. The first is pensions, which includes both the pensions awarded to academicians and the bonuses, wages, and relocation expenses paid to them and their assistants; this category cost 1,100,158 livres, or 51.4 percent of the total. The second is construction and maintenance of the Observatory, which cost 729,012 livres, or 34.1 percent of the total. The third is research, including supplies, equipment, retainers for maintenance of equipment, and engravings, which cost 310,098 livres, or 14.5 percent of the total. The average yearly cost was 62,920 livres.[1]

Taken as a whole, the seventeenth-century Academy's budget is comparable with the smallest annual income of "one of the richest monasteries in France" during the same period.[2] The abbey of Saint Germain des Prés

---

[1] Stroup (*Company*, chaps. 4–5) analyzes the finances of the seventeenth-century Academy.
[2] Ultee, *Abbey*, 65.

enjoyed from its endowment of land around Paris a yearly income of 60,000 to 100,000 livres. The abbey and the Academy were alike in depending on complex physical plants, having many members, and contributing to scholarship. Several academicians were lodged in the Academy's quarters, just as monks were housed in the abbey. Unlike the abbey the Academy had no endowment but depended on direct subsidy from the royal treasury, and the two institutions used their funds in very different ways. The abbey devoted 66 percent of its income, or 40,000 to 66,000 livres a year, to its buildings and grounds, 25 percent, or 15,000 to 25,000 livres a year, to such needs of its fifty-member community as food, clothing, and books, and 5 percent, or 3,000 to 5,000 livres a year, to "spiritual purposes and extraordinary expenses."[3] In contrast, only 34 percent of the Academy's funds went towards its physical plant, while 51.4 percent went towards pensions and 14.5 percent towards research (table 8).

The different budgets of the two institutions reflect their different characters. Thus the abbey had more money than the Academy but spent more on its physical plant and less on its members and their work than was the case for the Academy. The abbey sank the largest part of its income into the maintenance of its buildings and grounds. Spending on the Academy reflects no comparable responsibility once the Observatory was built, because maintenance of the Academy's physical plant was conflated with that of all the royal buildings. Monks and academicians had different circumstances which determined their aspirations and requirements. It is not surprising that the abbey spent less on each monk than the crown spent on each academician. Combining the disbursements for support of the community with those for spiritual and extraordinary purposes, the abbey spent at most 18,000 to 30,000 livres a year on the monks, in comparison to the 41,500 livres a year which the crown spent on pensions and research support for academicians. Yet while academicians, many of whom had to make a living and support families, had the greater financial need, it is not clear that they received proportionally more than the monks of Saint Germain des Prés; on the other hand, unlike the monks, some had additional sources of support.

Taking the seventeenth-century Academy as a whole, a comparison with the abbey of Saint Germain des Prés confirms the crown's generosity towards the fledgling scientific society. Receiving an average of 62,920 livres a year in known direct subventions, the Academy enjoyed an income comparable to that of a wealthy monastery. It also profited from being housed in various royal buildings and benefited from other indirect subsidies whose total extent remains unknown, as for example when its books were published by the Imprimerie royale.

But taking an average of annual expenditure over the full thirty-four year period can be misleading, for it underestimates spending before the 1680s and exaggerates spending thereafter. A summary of the average

---

[3] Ibid., 64, 96.

spent yearly by each minister (table 8) shows that only during Colbert's protectorship was the annual level of the Academy's expenditure comparable to that of Saint Germain des Prés. Indeed, while Colbert spent nearly 88,000 livres annually on the Academy, his successors spent less than half as much each year. Even if the construction and maintenance of the Observatory are subtracted from direct expenditure, Colbert's level of spending remains substantially higher than that of Louvois and Pontchartrain. Two general trends are clear from 1666 through 1699: direct expenditure declined, and the allocation of funds shifted under successive ministers, with the Observatory diminishing and pensions increasing as a proportion of the total budget.

Pontchartrain reinforced these trends. Moreover, he reduced the research budget, not only in absolute terms, as Louvois had done, but also in relative terms, as is clear from a comparison with the budget of another establishment funded by the crown, the Jardin royal. This was a teaching institution founded under Louis XIII to modernize medical training in Paris, and it offered lectures in botany, chemistry, and anatomy. It shared personnel and quarters with the Academy, but was the lesser organization in size and in scholarly importance. Its staff was smaller, its facilities less substantial, and its projects less ambitious than those of the Academy. From 1691 through 1698, when the Academy cost on average 34,173 livres a year, the Jardin royal cost on average about 19,000 livres a year, 55.6 percent of the Academy's cost.[4] Thus, the Academy was more expensive than the Jardin royal, but not proportionately so, given its size and responsibilities. Also, during this time of austerity, the Academy's expenses and pensions were reduced, while the costs of the Jardin royal were rising, perhaps because the teaching institution fulfilled its duties more assiduously than the research society, or perhaps due to the power of Fagon, who as physician to the king controlled the Jardin.[5]

Thus, the Academy cost proportionately less than either the Jardin royal or the abbey of Saint Germain des Prés, and by the 1690s it was a relatively inexpensive undertaking for its type. Even so, the crown spent more than three hundred thousand livres on the Academy during the 1690s, which it could have saved by suppressing its subsidies to the institution altogether. But cessation of patronage would have reduced the value of the Academy as propaganda for the reign. Instead of withdrawing support, the crown reduced expenditure on the Academy, while simultaneously organizing and funding a sister-society with more clearly defined practical responsibilities.

---

[4] For the expenses of the Jardin royal from 1690 to 1698, see *CdB*, 3: 425–27, 429, 437–38, 497, 565, 568, 570, 581–82, 642–43, 698–99, 714, 716, 718, 719, 722, 729–31, 792, 842, 848–50, 852, 863–64, 929, 982–83, 985–88, 991–92, 996–97, 1011–12, 1019, 1063–64, 1118–19, 1125, 1143, 1145, 1152, 1193; 4: 43, 46, 54–55, 72, 127, 150, 183, 185–86, 191, 193, 198, 210, 216–17, 272, 332–34, 344–45, 362, 417.

[5] On medical politics and the Jardin royal, see Howard, "Medical Politics," and Brygoo, "Médecins de Montpellier."

A literal reading of the treasury accounts from the 1690s suggests that Pontchartrain failed as protector of the Academy, for he could not sustain it or its members financially. However, while Pontchartrain financed research only at minimal levels and failed to safeguard the *practice* of pensioning academicians during these difficult years, he succeeded in guaranteeing the *rights* of academicians to their annual pensions. This was a significant accomplishment which presaged the permanent establishment of the institution by the *règlement* of 1699.

Fontenelle rightly praised Pontchartrain and the king for continuing support during the 1690s, but he exaggerated both the support and its effects. Contrary to his claims, payment of pensions *was* "interrupted during the great needs of the State," and the sciences *were* "injured by" the "cruel tempest" which "agitated all of Europe."[6] Lister was closer to the truth when, after listing the advantages of royal funding, he nevertheless concluded that "the war . . . has lain heavy upon the Philosophers too."[7]

Pontchartrain's protectorship during the 1690s was anomalous for the Academy. He strove for two inconsistent goals: improving the condition of the demoralized Academy and reducing expenditure on it. On the surface, Pontchartrain appears to have revived the Academy. With Bignon, the minister appointed new members, encouraged publication, and continued financial support during difficult years. As a result, the Academy was briefly invigorated, published several volumes, some of which had long been delayed, and was reorganized under an official *règlement* in 1699. But behind the scenes, funding was slow and severely reduced, pensions were rarely awarded to new members, and Bignon was constantly driven to justify even traditional expenses. Attendance at meetings was sporadic, and the members Pontchartrain himself appointed were among the worst offenders. By the end of the decade, fewer papers were presented in meetings, and there was very little long-range planning. In addition to these signs of deterioration, the Academy itself was changing as a result of its cumulated experiences, and its rooms at the Bibliothèque du roi were losing their importance as research facilities, for academicians worked at home or at other institutions.

In sum, during the 1690s Pontchartrain was more effective in restructuring the Academy and publishing its previous work than in stimulating new research. He substituted regularization and fiscal stringency for the generosity the Academy had originally enjoyed under Colbert. In these latter policies Pontchartrain continued a trend perceptible from the end of Colbert's regime and clearly visible in the eighteenth century. Thus the 1690s were both coda to the seventeenth century and prelude to the eighteenth.

---

[6] *Histoire . . . 1699*, 2 (Fontenelle's introduction to the *règlement* of 1699).
[7] Lister, *Journey*, 82.

APPENDIX A.

ON USING THE *COMPTES DU TRÉSOR ROYAL* AS A SOURCE OF INFORMATION ABOUT THE ACADEMY OF SCIENCES

From 20 September 1689 to 5 September 1699 Pontchartrain was *contrôleur général des finances* and therefore responsible for the Trésor royal. The records used to elucidate the finances of the Academy during the 1690s date, therefore, from his regime. As controller general, Pontchartrain sat on the *Conseil royal des finances*, conferred frequently with the king, and directed a sizable bureaucracy devoted to receiving and disbursing funds and recording these activities. Many of the records thus produced have been destroyed, but some survive in the Archives Nationales in the series G^7.[1] G^7 891–904 contain the summaries of receipt and expenditure for fiscal years 1690 through 1699. Associated documents for the same period include G^7 991–97, which contain some of the *estats* that initiated payment by the Trésor royal, along with requests to be included in the *état de distribution;* and G^7 980–87, which consist of ledgers recording the items authorized for the *états de distribution*.

These materials are crucial to the present analysis, but they are difficult to use and to cite. Although related to the correspondence of the controllers general edited by Boislisle in the nineteenth century, the accounts of the royal treasury have not themselves been published or indexed, and no guide exists to their use. Since most of their folios remain unnumbered, a description of their organization is necessary to explain the references to these documents. The following is offered, therefore, as a brief explanation of the contents and arrangement of the documents in AN G^7 891–904. The records for 1691 through 1698 (G^7 892–902) and for 1699 (G^7 903–904) will be discussed separately, because when Chamillart took over the controller generalship the nature of these records changed. This is not an exhaustive description, however, and it emphasizes the documents concerning expenditure.

The surviving records of receipt and expenditure of the Trésor royal during the 1690s (G^7 891–904) are stored in fourteen boxes. These are the

---

[1] *État de la France* (1692), 2: 277–78 and subsequent years under the title "Conseil Roïal des Finances"; *Correspondance des contrôleurs généraux;* Lavisse, *Louis XIV*, 2: 399–434; Mousnier, *Institutions*, 2: 154–57, 193–98. For an account of what was destroyed during the Revolution, including more than seven hundred volumes of documents from the seventeenth-century Trésor royal, see Viard, "Opérations du Bureau du triage."

papers of the *gardes du Trésor royal,* who had the responsibility for paying all the "dons et gratifications" made by the king. There were two such offices, and the two *gardes* served in alternate years. During the 1690s Nicolas de Fremont, Jean-Baptiste Brunet, Pierre Gruyn, and Jean de Tur-menyes held the offices, and their names appear on many of the folders in the boxes.[2]

The fourteen boxes are organized so that each contains the records of one fiscal year. The records of some fiscal years take up two boxes, and only G^7 904 contains information about more than one fiscal year. Since the crown had difficulty in raising money during the 1690s, the accounts for any one fiscal year extend over several calendar years.

Within each box are folders filled with folio-length pages; these may be single or in sewn or unsewn groups. Most of the folders are clearly labeled, and most boxes for the fiscal years 1690 through 1698 contain the following:

(A) *Revenue.* A large folder contains information about annual revenue, with a separate page or group of pages devoted to each type; this folder may account for one-third to one-half of the contents of a box.

(B) *Monthly Accounts.* More than a dozen folders record monthly income and disbursements, with a separate folder for each month during the first twelve months. In subsequent calendar years, as revenue and disbursements in the fiscal year continued, the monthly records became shorter and tended to be combined in quarterly or semiannual folders.

Additional information exists for some years. Only the data recorded in (B) were pertinent to this study.

The monthly accounts in (B) fall into two physically distinct categories: revenue (*la recette*) and disbursement (*la dépense* or *la despense*). Since only the latter shed light on the Academy, the following discusses only the arrangement of materials relevant to that category.

The monthly records of disbursement usually contain four kinds of information:

(1) *Somme Totalle de la Despense Actuelle.* This record of "total real expense" summarizes in a brief paragraph the total actual expenditure for the month and compares it with actual revenue for the month; it is written on a quarto-size page which sometimes serves as a folder for (2) through (4).

(2) *Despense Actuelle* (DA). This record of "real expense" lists only figures; its pages are keyed to registers now lost. It is missing altogether in G^7 902 and often does not exist for the late payments in a fiscal year.

(3) *Despense du Grand Comptant* (GC). This is usually a sewn group of pages and summarizes large expenditure, usually of one thousand livres or more. Entries take a standard form, starting with the preposition "To,"

---

[2] *État de la France* (1692), 2: 645–46 and subsequent years under the heading "Thrésor Royal."

that is, "A" or "Aux," giving the titles and names of the recipients, the purpose of the expenditure, the working year for which the recipient was paid, and the date of the *ordonnance* on which payment was based. The sum paid is listed in a column on the right-hand side of the page. At the bottom of each page is a cumulative total for the month.

(4) *Despense du Petit Comptant* (PC). This is usually a sewn group of pages and summarizes small expenditure, normally of less than one thousand livres. It presents the information in the same fashion as in (3).

In certain fiscal years, for example, 1694 and 1695, there are discrepancies in (B). That is, the total of (2), or actual as opposed to fictitious expense, does not correspond to the total of (3) and (4), the list of payments from the *grand* or *petit comptant*.[3]

Evidence of expenditure on the Academy from 1691 through 1698 has been found in (2), (3), and (4), and is summarized in tables 1 and 5. References in the present work to this evidence provide the number of the series (G[7]), the number of the box (892, 894, etc.) in which the document is stored, the month in which payment was recorded (Mar., June, etc.), and the designation "GC," "PC," or "DA" for *grand comptant, petit comptant,* or *despenses actuelles*.

The records for 1699 contained in the two boxes numbered G[7] 903 and G[7] 904 were prepared for a different purpose and hence are organized differently. Intended to provide Chamillart with a review of the state of the treasury when he took over the controller generalship from Pontchartrain, they include the following:

(C) *Summary, 1689-99* (G[7] 904). This summary of revenue, with the balance sheet of revenue and expenditure during the ten years of Pontchartrain's regime, is organized in separate folders, one for each fiscal year except 1698, with fuller data for some years than for others. A separate thin folder contains a summary of "Revenus ordinaires du Roy" from 1689 through 1697.

(D) *Revenue, 1699* (G[7] 903). Several folders summarize revenue, anticipated and actual, for the fiscal year.

(E) *Accounts Payable and Expenditure, 1699* (G[7] 903–904). Several folders summarize expenditure, anticipated and actual, for fiscal year 1699.

Information about the Academy occurs in the following documents in category (E):

(5) *Ordonnances à délivrer* (G[7] 904). This weekly summary of requests for payment provides the name of the person to whom the *ordonnance* was to be delivered, the name of the person entitled to the sum in ques-

---

[3] The formal distinction in monthly reports between real and anticipated expenditure or revenue was an innovation of 1684 and 1685 which allowed the treasury to balance the books each month: *Correspondance des contrôleurs généraux*, 1: 581–82.

tion, and the amount owed. As rudimentary as this information may seem, some items are unquestionably intended for the Academy. Thus, the names Fontenelle, Homberg, Couplet, and Simonneau—associated throughout the 1690s with the Academy of Sciences and the Compagnie des arts et métiers—usually occur together. All but three of these entries are for *ordonnances à délivrer* to M. Des Granges; the remainder cite *ordonnances à délivrer* to M. Couplet.

(6) *Reste des despenses 1699* (G^7 903). This folder contains (a) "État auquel . . . ," a two-page summary of revenue; (b) "Recette 1699"; and (c) a memorandum called "Despenses de l'année 1699 restant à payer au sixieme septembre 1699," consisting of sixty-nine numbered folios. The reference to the pensions owed to academicians occurs on f. 64 of (c). That reference is anticipatory: the *estat* for the Academy was not ordinarily prepared before the end of the year, and the *ordonnance* for the 1699 pensions was dated 26 January 1700. This document resembles that published by Esnault and informs Chamillart of the expenses he must anticipate for the remainder of this fiscal year.[4]

(7) *Despenses 1699 a arrester* (G^7 903). This folder contains (d) "Consommations 1699 arrestées" and (e) "Enregistremens des depenses 1699 a arrester par le Roy," dated "a Fontainebleau le 23 aoust 1712." These are unnumbered pages, keyed to other registers. "Enregistremens des despenses 1699" lists expenditure according to *chapitres*, that is, budget categories such as "Maisons Royalles," "Fortifications," and "Argenterie"; academicians' pensions were listed in the *chapitre* called "Gratiffications par comptant et autres despenses."

(8) *Despense Actuelle (DA)* (G^7 903). This is the equivalent of (2) and is recorded in a document entitled "Somme totale de la depense actuelle faite en mon Tresor Royal depuis le premier janvier 1700 jusqu'au premier mars 1708 pour l'exercice de 1699." The sums are listed so that payments from the *grand comptant* precede those from the *petit comptant*. This document confirms that academicians' pensions for 1699 were paid between January 1700 and March 1708, and that the *ordonnances à délivrer* from December 1699 through January 1700 for Couplet, Fontenelle, Marchant, and Homberg were paid. Since the sums earmarked for Couplet in *ordonnances à délivrer* before December 1699 in (5) cannot be correlated with the figures listed here, they must have been paid before January 1700.

Some records of the *grand* and *petit comptant* [see (3) and (4) above] are found in G^7 903 in a folder headed "M Gruyn 1699 depuis le 22 janvier 1701 jusqu'au dernier decembre 1707."

References to the evidence in AN G^7 903–904 provide the number of the series, the number of the box, the date of an *ordonnance*, and sometimes

---

[4] Esnault and Boislisle, "Documents relatifs," and [Chamillart], "État."

the title of the folder. The descriptions of (5), (6), (7), and (8), above, provide more detailed information.

The surviving archival sources provide clues to the bureaucratic practices of the Trésor royal and indicate that any request for funds had to go through several steps before an individual actually received payment. Two documents in the *Correspondance des contrôleurs généraux*[5] summarize the procedures established by Colbert and le Peletier. The surviving accounts of the treasury during the 1690s are consistent with the procedures described there. But the two documents emphasize the treasury's record-keeping and accountability to the king rather than any methods for obtaining payment. Since inferences about all of these procedures affect the present analysis, it is important to explain some of them briefly. The following is not an exhaustive summary, but rather a circumstantial account which balances the information in the *Correspondance des contrôleurs généraux* against a reading of the accounts of the treasury pertinent to the Academy; it stresses the procedures for collecting monies from the treasury.

The first step towards payment occurred when the crown agreed in principle that a group or an individual was entitled to receive funds from the royal treasury. Such funds usually represented payments for services rendered; and they included stipends for the holders of purchased offices, reimbursements for purchases or expenses incurred on behalf of the crown, *gratifications* to scholars, salaries for professors at the Collège royal, and so on. Colbert established this right for the Academy of Sciences when he founded it in 1666, and Pontchartrain preserved that right in the 1690s, as the records of the Trésor royal demonstrate. The crown's disbursements during the 1690s are similar from year to year, and most of the surviving petitions for payment were requests for ordinary entitlements rather than for new support.

Academicians received two kinds of payment for their work at the Academy of Sciences: pensions and reimbursements. To receive either, it was necessary to take the second step and submit a request, which in the case of pensions was called an *estat*. La Chapelle had the responsibility of drawing up the *estat* for pensions when Pontchartrain first became protector of the Academy, but Bignon later assumed that task. Several *estats* for pensions survive.[6] They summarized the claims of members of the Academy of Sciences, the Academy of Inscriptions, and the two translators. Each *estat*

[5] *Correspondance des contrôleurs généraux*, 1: 578–82 (app. XIII), from AN KK 355. AN KK 355 (13v, 332r) notes that Pontchartrain changed none of the procedures established by Colbert and modified by le Peletier.

[6] Although seventeenth-century orthography used "état" and "estat" interchangeably, I have for the sake of clarity used "estat" to refer to these manuscript lists of academicians and their pensions, and "état" for all other meanings of the word. AN G[7] 992, 973, O[1] 656, 1934[B] 14; table 1 summarizes their contents. The *estat* for 1712 also survives in AN G[7] 973, and BN MS. fr. 22225 contains *estats* from 1707. Bourdelin's notebooks provide evidence about the costs of drawing up requests for reimbursement for his expenditures on the chemical laboratory: BN MS. n. a. fr. 5147, 5149.

reported the working year for which pensions were owed, listed the academicians who were entitled to pensions, and summarized the work they had done in the Academy that year; the amount due to each was written in both prose and numerals, and the total due on the *estat* was stated at the end. Sometimes an old *estat* was revised by crossing off the previous date and replacing it with that of the new working year. Once completed, the *estat* was presented to Pontchartrain as *secrétaire d'état* with responsibility for the Academy; he initiated the procedures within the royal bureaucracy which led to payment.[7]

The third major step was approval of the *estat* for payment. This was supposed to take place in the Tuesday and Saturday meetings of the *Conseil royal*, whose business included a review of *estats* and *ordonnances* presented to the treasury. The king alone had the right to authorize payment of the sums requested, and he wrote "bon" or "payez" and signed "Louis."[8] Once the *estats* and *ordonnances* were signed, the controller general handed them over to a *commis*, who recorded them according to category, or *chapitre*, and by date.[9] That date served to identify an *ordonnance* in the future. Thus many of the entries concerning academicians refer to an *ordonnance* of a particular date. A characteristic example is this entry for their pensions: "Aux gens de lettres gratification 1690 suivant l'estat et ordonnance du 21 janvier 92. . . . . 36000."[10]

The fourth step, like the second, required initiative by the academician, who had to campaign to have his name included on the *état de distribution*. That was the list, prepared by the controller general and reviewed by the king, of those who could get cash payment. This step was necessary only for items payable from the *grand comptant*.[11] The documents in AN G[7] 991–97, some of which are transcribed in appendix B, bear witness to the perseverance, servility, and desperation of many persons who struggled to have their *ordonnances* placed on the *état de distribution*.[12] Success could take a long time, as is clear from the contents of the surviving petitions, as well as from the lapse of time between the date of an *ordonnance* and

---

[7] *Correspondance des contrôleurs généraux*, 1: 578–80.

[8] Mousnier, *Institution*, 2: 156; Antoine, *Conseil*, 49, 58–60; Boislisle, "Conseils," 6: 498–500, 507–10.

[9] De Bie kept the *registres* of "fonds et dépenses" under le Peletier and Pontchartrain: AN KK 355: 332r.

[10] AN G[7] 894, July 1692, GC. The dates of only three of the *ordonnances* governing the pensions for the Academies fell on Tuesday or Saturday; thus most of these dates do not correspond to the twice weekly meetings of the *Conseil royal*. No date is known for the *ordonnance* governing pensions due for 1695.

[11] *Correspondance des contrôleurs généraux*, 1: 580. In the eighteenth century, Pontchartrain informed Desmaretz that the king reserved for himself the right to sign the *états de distribution*: Mousnier, *Institution*, 2: 156.

[12] Approval could take the form of "bon" written on a petition (see appendix B) or of the note "A mettre sur l'etat de distribution" written on a slip of paper bearing the names of the fortunate; such a note includes the name of Charpentier, a member of the Academy of Inscriptions, in AN G[7] 993, 22 Feb. 1695. After an item was approved for inclusion on the *état de distribution*, Pontchartrain turned the document over to a *commis*, for example, De Bie or Vallée.

its inclusion in the *paiements ordonnés* (AN G⁷ 980–87), which correspond to the *état de distribution*.

The fifth step was the payment of funds by a *garde du Trésor royal* either to a treasurer or to those whose names were on the *état de distribution*. The date of payment was usually very shortly after the date when an item was recorded in the ledgers called *paiements ordonnés*. Ideally, this would be a cash payment equal in value to the amount owed in livres, deniers, and sous, which were units of account.

The treasury adopted complicated mechanisms for settling some of its debts. For example, many *estats* or *ordonnances* were approved as *assignations*; that is, the crown promised to pay the sum out of specific revenues designated by it. In 1697 Racine and Boileau Despréaux asked that their pensions for 1695 as royal historiographers be paid on this basis.[13] But another solution was found for the Academies of Sciences and Inscriptions and for the professors royal. Payment of their pensions became in certain years a partial fiction. Although academicians did not receive the cash amount of the pensions listed on the *estat*, the crown was entitled to show on its books that it had settled its obligation. This was accomplished by turning the pensions into *rentes*, that is, into investments which obliged the treasury to pay each year a percentage of the principal as annuities to the academician. This arrangement was better for an academician than receiving nothing at all, but it was very close to being a forced loan to the crown.[14]

The record of these transactions survives piecemeal in G⁷ 994, 995, and 996. It shows that academicians requested *rentes* in place of their pensions when they had not received pensions for three or more consecutive years. Because only some of the supporting documents exist, it is not possible to ascertain how many academicians took *rentes* instead of pensions or for how many years they did so. The abbé Tallemant, however, believed that most members of the Academy of Sciences had done so, and he expected most members of the Academy of Inscriptions to do the same.[15]

The transformation of pensions into *rentes* during the mid-1690s clouds any analysis of Pontchartrain's protectorship of the Academy. Without knowing who received pensions, who received annuities, how many years' worth of pensions were transformed into annuities, what the size of each academician's annuity was, or whether the annuities were paid regularly and for the duration, it is impossible to establish the precise cash cost to the crown of pensioning academicians during the 1690s. In order to compare Pontchartrain's regime with those of Colbert and Louvois, therefore, the *budget* for, rather than the *cost* of, pensions has been taken as the common measure.

---

[13] AN G⁷ 997, 12 Feb. 1697. In 1696 it became increasingly common for those who were owed pensions to request *assignations* for their payment: AN G⁷ 996.

[14] See chap. 3.

[15] Appendix B, document IV.

In summary, the records of the royal treasury during the 1690s contain references to the Academy of Sciences and its members which make it clear that academicians were entitled to pensions throughout the decade, that they received some of their pensions in toto and the rest as annuities, and that several academicians were reimbursed for research expenses. Some of the more important documents have been transcribed in appendix B, and the details culled from the archival record have been summarized in tables 1, 2, 3, and 5.

Most studies of the Trésor royal have emphasized revenue rather than expenditure. When expenditure has entered the picture, it has primarily been used as an index of the parlous state of royal finances, by comparing it with revenue in light of the crown's growing indebtedness. Moreover, analyses of royal expenditure have tended to concentrate on the grandiose examples, especially the enormous outlays for Versailles and other royal palaces, and to discuss them in the context of the failure of revenue to cover expenses. In the eighteenth and nineteenth centuries,[16] therefore, when Louis's expenditure was discussed at all, it was often presented as excessive, frivolous, and deleterious to the health of the treasury and the kingdom. Although Guiffrey published the *comptes des bâtiments* and drew attention to the variety and depth of information in records of royal disbursement, these and similar documents continue to be neglected.

The accounts of the Trésor royal offer more to the historian, however, than spectacles of insolvency or profligacy. The records of both income and expenditure are a rich source of information about the reign, and the latter clarify Louis XIV's allocation of increasingly scarce resources. The documents concerning *dépenses* shed light not only on the learned societies, but also on pensions to individuals such as Madeleine de Scudéry, to new converts from Judaism and Protestantism, to engineers, or to the royal musicians, the "comédiens Italiens," and "les danseurs des ballets." In them are recorded the monthly subventions to the king of England, the amounts spent on "affaires secrettes," payments to ambassadors, and other significant details. It is unrealistic to hope that these documents will be published, but they should be recognized more widely as a valuable source of information about the reign of Louis XIV. Difficult as they are to exploit, they merit further attention from historians and archivists.

---

[16] Forbonnais, *Recherches;* Malet, *Comptes rendus;* Bailly, *Histoire financière;* Chéruel, *Histoire de l'administration;* Rousselot de Surgy, "Discours préliminaire"; BN MS. fr. 7750: 1–41, Pegere's "Abregé historique sur les finances."

APPENDIX B.

DOCUMENTS FROM THE RECORDS OF THE TRÉSOR ROYAL WHICH SHED LIGHT ON THE ACADEMY OF SCIENCES

I. In this document, Jacques de Tourreil, a member of the Academy of Inscriptions, requests payment of his pension accumulated for the four years from 1691 through 1694 in the form of an annuity. It is odd that he sought four years' worth of pensions, because he was included on the *estat* for 1691, which was paid in April 1693 (table 1, fiscal year 1692), and later got approval for a cash payment of his 1695 pension in the spring of 1696 (AN G^7 996, between 21 Mar. and 29 May 1696). Here as in the other documents transcribed, the marginalia and "bon . . ." at the head or foot of the document, represent the controller general's recommendations and the king's approval.

Monsieur de Bie

> Le Sieur de Tourreil est sur l'estat des gens de lettres pour deux mils francs de pension annuelle, les arrerages de sa pension accumulés depuis mil six cents quatre vingt onze montent avec l'année courante a huit mil frans. Il demande a les convertir en une rente, dont il puisse tirer une partie des secours qui luy sont absolument necessaires pour attendre, que les liberalitez du Roy reprenent leur cours ordinaire.

bon en rentes viageres

AN G^7 994, 5 Apr. 1695

* * *

II. Philippe de La Hire asks to transform his pension as academician and his *gages* and *augmentations* as professor at the Collège royal into *rentes au denier 18*, that is, annuities equal in value to one-eighteenth of the principal invested. By increasing the principal to 7,200 livres, the crown made it possible for La Hire to receive an annuity of 400 livres. Compare document VI, according to which Du Verney had to increase the principal out of his own funds.

> [r] Le Sieur de La Hire demande que les sommes a luy ordonnées par Sa Majesté pour les années 1692, 1693, et 1694, par les etats des Professeurs Royaux, et de l'Academie Royalle des Sciences soient converties à son profit en rentes sur l'Hostel de Ville.

Il luy est deu comme Professeur es Mathematiques pour ses apointemens de l'année 1692 par ordonnance du 23 mars 1693 la somme de 600ᵘ

Et par augmentation 200

Plus pour ses apointemens de l'année 1693 par ordonnance du 25 janvier 1694 600

Et par augmentation 200

Plus pour ses apointemens de l'année 1694 par ordonnance du 22 janvier 1695 600

Et par augmentation .... <u>300</u>

2500ᵘ

[v] De l'autre part .... <u>2500ᵘ</u>

Plus a luy par gratification comme astronome de l'Academie Royalle des Sciences, pour l'année 1692 par ordonnance du 16 fevrier 1694 1500

Plus par autre gratification pour l'année 1693 par ordonnance du 28 avril 1694 1500

Plus par autre gratification pour l'année 1694 par ordonnance du .... 1695 .... <u>1500</u>

Total .... <u>7000ᵘ</u>

7200ᵘ bon au denier 18 en adioutant le quart.

The abbé Jean Gallois also requested that his pensions accumulated from 1692 through 1694 be transformed into *rentes*. His petition followed the format of La Hire's and it was approved in the same week as La Hire's. Gallois collected 600 livres in *gages* and 100 livres in *augmentations* as professor of Greek at the Collège royal; 1,000 livres for his *gratification* as "préposé a la direction du College Royal," that is, as overseer of the Collège royal; and 1,500 livres for his *gratification* as former secretary of the Academy of Sciences. Since he was due these sums for 1692, 1693, and 1694, the total owed him was 9,700 livres.

The crown did not increase Gallois's principal as it did for La Hire, but Gallois's contract was approved for *rentes au denier 18 en doublant*. This was often done for academicians who invested in *rentes viagères*, which lapsed on the death of the individual and, thus, could not be inherited. The rate was the equivalent of one-ninth of the principal. Since Gallois's principal was 9,700 livres, his annuity would have been 1,078 livres.

Similar contracts exist for other professors royal. Enguehard's was approved "en rentes au denier 18 adioutant 1500 livres," giving him a principal of 3,600 livres. Gerbais's accumulated salary was worth 3,600 livres, Cappelain's worth 3,200 livres; both were awarded *rentes* "au denier 18 en doublant."

AN G⁷ 994, 24 May 1695

* * *

III. Joseph Sauveur requests payment of his salary as professor royal. The official "bon," approving the request, appears in the upper left-hand corner of the sheet with the name of Pontchartrain's *commis*; at the bottom is a notation of Sauveur's *gages* of 600 livres and his *augmentations* of 100 livres. Note that Sauveur was not yet an academician (he was appointed 29 Feb. 1696) and was apparently not known to Pontchartrain. The "cartes marines" which exhausted Sauveur's funds were published as the *Neptune françois* and are discussed in chap. 6.

A Monseigneur de Pontchartrain

Monsieur de Bie
<u>bon</u>

> Monseigneur
> Le depart de Monseigneur de Phelipeaux avec qui jay eu l'honneur de travailler deux ans et qui auroit eu la bonté de vous parler en ma faveur, m'oblige, Monseigneur de vous suplier tres humblement d'ordonner le payment de ce qui mest du de mes apointemens du College Royal, n'ayant aucun moyen de subsister et m'estant epuisé d'ailleurs par limpression des cartes marines.

<u>Sauveur</u>

100
600-

AN G⁷ 994, 31 May 1695

* * *

IV. The abbé Paul Tallemant, secretary of the Academy of Inscriptions, petitions to transform his pension into *rentes viagères* and suggests that other members of the Academy of Inscriptions be given the same opportunity. At the end of the document Tallement explains how his resources are being exhausted by the exceptional taxation of the decade. The academician requested either of two alternatives for himself, namely, that he receive cash payment for one of his pensions or that he use both of them to acquire *rentes viagères*. The "bon" approving Tallement's first alternative has been crossed off and replaced by a "bon" approving the second instead. At the time that Tallement made this request, the crown had not increased his pension by 1,000 livres, as it was to do in the *estat* for 1695. Tallement corrected his statement in three places, crossing out words and replacing them with others; this has been indicated in the transcription.

[r] A Monseigneur de Pontchartrain
> L'abbé Tallement a esté payé d'une année de sa pension pour 1692 par la bonté de Monseigneur de Pontchartrain.

> Il a remarqué que ~~tout~~ la pluspart de Messieurs de l'Academie des Sciences et les Professeurs Royaux, ont pris des rentes

viageres p̶̶̶̶̶̶ ou des rentes au denier 18 en doublan moyennant
quoy ils ont le payement des trois années échûës, scavoir 92, 93,
et 94.

Si on avoit sçû que Monseigneur de P. eust eu cette invention
ou pour mieux dire cette bonté, Messieurs de l'Academie des In-

[v]   scriptions l'auroient // suplié de leur faire la mesme grace; et
l'abbé Tallemant en parlera a ces Messieurs si Monseigneur le
souhaite ainsy parceque cela nettoyeroit entierement les estats des
trois années, et il ne doubte pas qu'au moins pour 93 et 94 ils
n'acceptent volontiers le parti.

A l'egard de l'abbé Tallemant comme il avoit quelques petits
engagemens pour l'année 1692 il n'a pû se dispenser de toucher
les 1500ᵘ de ladite année.

C'est pourquoy il suplie tres humblement Monseigneur ou de luy
b̶o̶n̶   faire encore payer l'année 1693 ou de luy permettre d̶̶̶̶ d'ac-
[r]   querir pour les deux années restantes montant a // trois mille
bon   livres, un contrat de rentes viageres soûs le nom d'un de ses amis
qui luy compteroit cette somme.

Cela contribuëroit a le faire subsister cette année, dans laquelle
le franc alleu, les fontaines, la capitation, et le don gratuit épui-
seront infailliblement le revenu de ses benefices, et il joindroit
cette grace a tant d'autres dont Monseigneur l'a honoré.

AN G⁷ 994, 28 June 1695

* * *

V. Jean Dominique Cassini requests cash payment of his pension from
1692 through 1695. Only payment for 1692 was approved; when Cassini
actually received his pension, it was recorded as for 1693 (table 1, fiscal
year 1694).

M Cassini de l'Academie Royale des Siences demande le payement
des sommes qui lui sont ordonées sur les etats des gens de letres.

Il lui est ordoné par lesdits etats.
bon   pour 1692 9000
~~pour 1693~~ . . . . ~~9000~~
~~pour 1694~~ . . . . ~~9000~~
~~pour 1695~~ . . . . ~~9000~~:

AN G⁷ 995, 24 January 1696

* * *

VI. Joseph Guichard Du Verney petitions to transform his pensions from
1693 through 1695 into investments in the debt of the Hôtel de Ville. He

will add 900 livres to the sum owed him, bringing the principal to 7,200 livres. Since annuities on those investments were being calculated at denier 18, or one-eighteenth of the value of the principal, Du Verney would earn 400 livres a year. Note that *augmentations de gages*, such as the professors royal received, were already treated not as outright cash payments but as principal from which their recipient collected annuities. They provided a precedent for Pontchartrain's transformation of academicians' pensions into *rentes*.

> Il est dû a M. du Verney—
> Pour l'année 1693 .... 2100^u
> Pour 1694 2100
> Pour 1695 2100
> Total   6300
>
> bon$\qquad$Il demande d'être payé de cete some en augmentations de gages, sans être obligé de fournir pour suplement que la somme de 900^u qui luy feroit un contrat de 400^u de rente.

$\hfill$ AN G^7 996, document 125

*   *   *

VII. Cassini and the other astronomers in the Academy were allowed to start work again on the extension of the meridian in 1700. Two petitions survive from 1701 in connection with that project. The first requests payment of a special *gratification* of 3,000 livres for Cassini. This was authorized in December 1700 and was delayed for three months. But as the request pointed out, Cassini was on the French border, and so he could not collect the sum anyway.

> Le Sieur de Cassini gratification
> ordonnance decembre 1700 $\hfill$ <u>3000^u</u>
>
> On a expedié pour Monsieur Cassini une ordonnance de 3000^u mais depuis le 22 decembre elle na pas encor este mise sur lestat de distribution.
> Cependant Monsieur Cassini est sur les confins de la France.
> du 7 mars 1701. $\hfill$ a ~~lordinaire~~ a la fin de ce mois

$\hfill$ AN G^7 998 (1701)

*   *   *

VIII. The second document pertinent to the extension of the meridian represents reimbursements to Cassini and others for expenses incurred in carrying out this project. It also includes *gratifications* of 900 livres apiece to three academicians—Couplet, Maraldi, and Chazelles—assisting Cassini. This document, authorizing payment of 3,427 livres 7 sous, refers to another, authorizing an additional 6,000 livres' worth of payments. The total cost of the work, once this *ordonnance* was paid, would have been 9,427 livres 7 sous, plus Cassini's 3,000-livre *gratification*. Given the expense, it is not surprising that the project was delayed during the 1680s and 1690s. Note that "30 avril" appears in the margin beside the line containing the date "3ᵉ avril" and has been taken as a correction of the earlier date. This is a copy of the *ordonnance*, and the last sentence lists the signatures of approval.

<u>Au Sieur Cassinj 3427-7.</u>

Garde de mon Tresor Royal

Monsieur Pierre Gruyn payez comptant au sieur de Cassiny la somme de trois mil quatre cent vingt sept livres sept sols pour avec celle de vi^{g} ^{u} dont a esté fait fonds par ordonnance des 13 septembre et 13 avril derniers, faire la somme de ix^{g}iiii^{c}xxvy^{u} vy^{s} pour le parfait payement de la depence qui a esté faite tant par luy que par plusieurs mathematiciens pendant le voiage qu'ils ont fait dans plusieurs provinces de mon royaume pour des observations depuis le 20 aoust 1700 iusqu'au 3ᵉ avril [30 avril] de la presente année y compris ix^{c} ^{u} que i'ay accordé par gratiffication aux Sieurs Couplet, Maraldy et Chazelles qui l'ont accompagné audit voiage. Fait a Marly le 8 Juin <u>1701</u> signé Louis a costé comptant au Tresor Royal plus bas bon Louis et au bas Phelippeaux.

AN G⁷ 998 (1701)

APPENDIX C.

ESTIMATING THE PENSIONS PONTCHARTRAIN PAID TO MEMBERS OF THE ACADEMY OF SCIENCES

Pensions for members of the Academy of Sciences, the Academy of Inscriptions, and two translators were submitted together, on one *estat*, to the treasury. From 1666 through 1690 the treasury paid the amount requested to the *comptes des bâtiments*, whose records show the disbursements to each academician, along with the justification for and the date of payment. Under Colbert and Louvois, therefore, it is possible to chart the annual payments to specific academicians and to spot modifications of personnel and pensions from year to year. From 1690, however, the records are less forthcoming. The Academies were no longer under the purview of the buildings account, whose guardians kept detailed records. The accounts of the treasury during the 1690s simply list the total due on the *estat* for the two Academies, plus a few dozen payments authorized to individual academicians; only one *estat* from the decade has been found. Thus the amount intended for each Academy or for each academician, like the size of the average pension paid, can only be estimated on the basis of partial information.

The share of the *estat* that went to the Academy of Sciences can to some extent be inferred from the following evidence: the *estats* of 1690 and 1703, the amounts paid to other persons on the *estat* of the Academies, the list of *pensionnaires* in the Academy of Sciences for 1699, the total cost of the *estats* from 1689 through 1703, and the facts that Pothenot was excluded from the Academy of Sciences by 1699, that Des Billettes and Jaugeon continued to receive as members of the Academy of Sciences the same pensions they had collected as members of the Compagnie des arts et métiers, and that the *estat* of the Academies grew in 1698 by the size of Fontenelle's eventual pension. This information is summarized in table 1.

The implications of this information for establishing how the *estats* were divided are summarized in figure C.1. Starting with the annual total of the *estats* of the Academies, and subtracting the share that went to the translators, the assistants, and the Academy of Inscriptions, it is possible to calculate how much went to the Academy of Sciences.

Once Pontchartrain made up the arrears accumulated under Louvois, the amounts due on the *estat* sometimes remained unchanged for three consecutive years. For the working years 1692 through 1694 the crown budgeted 43,000 livres a year; from 1695 through 1697, 44,000 livres a

Fiscal Year	Working Year	Amount Due on *Estat*			ARdI Share			Translators' Share	Assistants' Share			ARdS Share
		lv.	s.	d.	lv.	s.	d.	lv.	lv.	s.	d.	lv.
1691	1689	23,933.	6.	8	6,466.	13.	4	1,600	1,066.	13.	4	14,800
1692	1690	36,000			9,700			2,400	1,600			22,300
1692	1691	39,900			13,700			2,400	1,600			22,200
1694	1692	43,000			13,700			2,400	1,600			25,300
1694	1693	43,000			13,700			2,400	1,600			25,300
1695	1694	43,000			13,700			2,400	1,600			25,300
1696	1695	44,000			14,700			2,400	1,600			25,300
1697	1696	44,000			14,700			2,400	1,600			25,300
1697	1697	44,000			14,700			2,400	1,600			25,300
1698	1698	45,500			14,700			2,400	1,600			26,800
1699	1699	47,800			15,000			2,400	1,000			29,400
	Total	454,133.	6.	8	144,766.	13.	4	25,600	16,466.	13.	4	267,300

FIG. C.1. The Share of the *Estat* That Went to Members of the Academy of Sciences, 1691–99 (Summary of Table 1 as interpreted in appendix C.)

year; in 1698, the total rose to 45,500 livres; and in 1699, it went up again to 47,800 livres. The two translators, Dippy and De La Croix, received 1,200 livres a year. The assistants to the Academy of Sciences got 1,600 livres a year until 1699, when Dalesme became a member of the Academy and his pension of 600 livres was included with those of other academicians. The rest of the *estat* was split between the Academy of Sciences and the Academy of Inscriptions. It is necessary, therefore, to ascertain how much each Academy received and which Academy was responsible for the increases in the amount of the *estat* in 1695, 1698, and 1699.

The Academy of Inscriptions, whose small size earned it the name "la petite compagnie," had a more stable membership than did the Academy of Sciences. From 1691 through 1700 there were only eight members at any one time. Their pensions are known from the *estat* of 1690 and its marginalia, from BN MS. Clairambault 566, and from the individual payments recorded in the accounts of the treasury. Pensions for the Academy of Inscriptions probably totaled a uniform 13,700 livres for the working years 1691 through 1694. Tallemant's pension rose by 1,000 livres in working year 1695, increasing the share of the Academy of Inscriptions to 14,700 livres through 1698.

In 1698 the *estat* of the Academies rose by 1,500 livres. Scrutiny of the membership of the two Academies and of information about subsequent distribution of pensions suggests that this sum was probably intended for Fontenelle, who became secretary of the Academy of Sciences in 1697 and who received a pension of 1,500 livres in 1703.

In 1699 the *estat* of the Academies rose again, by 2,300 livres. This was the year when the Academy of Sciences received its *règlement*, which included an understanding that henceforth twenty members would be *pensionnaires*. During the early eighteenth century, they divided 30,000 livres among them. To accomplish this, the crown increased the *estat* of the Academies. All but 300 livres of the newly available funds went to the Academy of Sciences. That is, when Jaugeon and Des Billettes joined the Academy, they continued to receive the 1,000-livre pensions which they had been earning in the Compagnie des arts et métiers. Dalesme, likewise, had received 600 livres a year as an assistant to the Academy, and this was the amount of his pension when he became a member. The 300 livres for the Academy of Inscriptions probably went to Dacier, for that was his basic pension in 1703. A bookkeeping device ensured that the Academy of Sciences actually collected more than the 30,000 livres allotted to it, for Chastillon's name was now listed on the *estat* with the Academy of Inscriptions. Thus, in 1699, the Academy of Sciences received 29,400 livres for its members, and 1,000 livres for its assistants, totaling 30,400 livres on the *estat*, 400 livres more than the 30,000 established by the crown (table 1, fiscal year 1699).

If this reasoning is correct, then it is possible to calculate the amount the crown budgeted on the *estat* for pensions on behalf of the Academy of Sciences from fiscal year 1691 through fiscal year 1699. As summarized in figure C.1, the *estats* for the eleven working years covered by this period totaled 454,133 livres 6 sous 8 deniers. Of that, 144,766 livres 13 sous 4 deniers were for the Academy of Inscriptions, 25,600 livres for the translators, 16,466 livres 13 sous 4 deniers for the assistants to the Academy of Sciences, and 267,300 livres for members of the Academy of Sciences. In addition, the crown paid 2,250 livres to two academicians off the *estat* (table 2). Members of the Academy of Sciences thus earned 269,550 livres. Adding the earnings of their assistants, all *pensions* and *gratifications* for the Academy of Sciences totaled 286,016 livres 13 sous 4 deniers (table 2). During the same period, pensions for the Academy of Inscriptions totaled 144,766 livres 13 sous 4 deniers. The combined total for the two Academies was 430,783 livres 6 sous 8 deniers. The Academy of Sciences accounted for 66.4 percent of that total, but it had on average 27.6 members (fig. 2.1) and three assistants from 1691 through 1698, as against eight members in the Academy of Inscriptions. Thus an Academy's share of the *estat* was not proportionate to its size.

Having described how the *estats* were probably divided between the two Academies, we can turn to the cloudier issue of how the sum destined for the Academy of Sciences was actually divided among its members.

Several members were not included on the *estat* as a matter of policy. Bignon, L'Hospital, Chazelles, Guglielmini, Lagny, and Langlade were appointed as president, honorary, associate, foreign, or external members and were excluded by definition from the right to a pension. La Chapelle and Thévenot were pensioned not as members of the Academy of Sciences but for their positions in the Academy of Inscriptions and the Bibliothèque du roi, respectively. Carré, Cassini II, P. Couplet, G.-P. de La Hire, Maraldi,

Fiscal Year	Working Year	To All Academicians	Total / Average a Year	To Huygens & Cassini	Total / Average a Year	Remainder	Total / Average a Year	Number Pensioned With Huygens & Cassini	Number Pensioned Without Huygens & Cassini	Average Pension With Huygens & Cassini	Average Pension Without Huygens & Cassini
		lv.	lv.	lv.	lv.	lv.	lv.			lv.	lv.
1666	1666	24,100		5,000		19,100		15	14		
1667	1667	29,900		6,000		23,900		19	18		
1668	1668	34,400		9,000		25,400		19	17		
1669	1669	38,050		12,750		25,300		21	19		
1670	1670	37,600		15,000		22,600		19	17		
1671	1671	42,225		15,000		27,225		21	19		
1672	1672	42,800		16,500		26,300		21	19		
1673	1673	40,100		15,000		25,100		20	18		
1674	1674	38,950		15,000		23,950		19	17		
1675	1675	37,950		15,000		22,950		17	15		
1676	1676	34,200		12,000		22,200		17	15		
1677	1677	32,100		9,000		23,100		16	15		
1678	1678	37,600		15,000		22,600		16	14		
1679	1679	34,800		12,000		22,800		16	14		
1680	1680	40,000		15,000		25,000		16	14		
1681	1681	33,100		13,500		19,600		15	13		
1682	1682	28,700		9,000		19,700		14	13		
1683	1682–83	7,950		6,750		1,200		2	1		
1666–83		614,525 / 34,140		216,500 / 12,028		398,025 / 22,113		16.83	15.11	2,029	1,463

1684	1683	21,600		4,500		17,100		14	13		
1685	1684	27,700		9,000		18,700		15	14		
1686	1685	26,200		9,000		17,200		17	16		
1687	1686	26,200		9,000		17,200		17	16		
1688	1687	26,200		9,000		17,200		17	16		
1689	1688	24,200		9,000		15,200		16	15		
1690	1689	7,400		3,000		4,400		15	14		
	1684–90		159,500/22,786		52,500/7,500		107,000/15,286	15.86	14.86	1,437	1,029
1691	1689	14,800		6,000		8,800		15	14		
1692	1690	22,300		9,000		13,300		15	14		
1692	1691	22,200		9,000		13,200		15	14		
1694	1692	25,300		9,000		16,300		17	16		
1694	1693	25,300		9,000		16,300		16	15		
1695	1694	25,300		9,000		16,300		16	15		
1696	1695	25,300		9,000		16,300		16	15		
1697	1696	25,600		9,000		16,600		16	15		
1697	1697	25,300		9,000		16,300		16	15		
1698	1698	27,300		9,000		18,300		17	16		
1699	1699	30,850		9,000		21,850		21	20		
	1691–99		269,550/24,505		96,000/8,727		173,550/15,777	16.36	15.36	1,498	1,027
Total, 1666–99			1,043,575/30,693		365,000/10,735		678,575/19,958	17.47	16.03	1,757	1,245

Note: This information includes pensions, supplements, and moving expenses paid to members of the Academy of Sciences. It excludes any such payments to the Academy's assistants. It is based on table 1, *CdB*, and Colbert, *Lettres*, 5: 466–98. The overall average is higher than the individual averages because of ministerial overlap for the working years 1683 and 1689.

FIG. C.2. Average Pensions Paid to Members of the Academy of Sciences, 1666–99

and Tauvry most likely went without pensions because they were students. Finally, it is reasonable to infer from the financial record that La Coudraye also was not pensioned: since Varignon took over Sédileau's pension after April 1693, unless someone else lost a pension when La Coudraye entered that year, no surplus existed for his pension. In any case, La Coudraye's exceptionally high rate of absenteeism (table 4) is inconsistent with receipt of a pension.

Besides Homberg and Tournefort, therefore, only five of Pontchartrain's appointments seem to have been candidates for pensions during the 1690s: Boulduc, Charas, Fontenelle, Morin, and Sauveur. But what we know about the *estat* of the Academies suggests that before 1698 only 100 livres were left over after Cassini, Homberg, Bourdelin, Dodart, Du Hamel, Du Verney, Gallois, La Hire, Tournefort, Marchant, Varignon, Méry, C.-A. Couplet, Sédileau, Pothenot, Rolle, Cusset, and Le Febvre were paid. When Cusset ceased to be paid, Varignon collected the amount of his pension, and when Sédileau died, his pension was also assigned to Varignon. The extra 100 livres, available for working years 1690 and 1692 through 1698, probably went to Le Febvre, as explained in table 1, fiscal year 1692, note b. Only after Pothenot was excluded did another 400 livres become available. The precise date of his exclusion is not known; judging from the record of attendance, it was probably in 1696 (table 4; table 1, fiscal year 1697). In 1698, 1,500 livres were added to the *estat,* probably to pension Fontenelle, who had taken over Du Hamel's duties as secretary. From 1691 through 1697, therefore, at most 400 to 500 livres became available for Boulduc, Charas, Morin, or Sauveur. Since Boulduc was a *pensionnaire* in 1699, when the *estat* was inadequate to pay pensions at the higher 1703 rates to Méry, Rolle, Varignon, and Couplet, the funds made available by Pothenot's exclusion were probably awarded to Boulduc (table 1, fiscal year 1699).

In summary, in 1691 and 1692 Pontchartrain paid the fifteen members due pensions for their work under Louvois. From 1692 Tournefort and Homberg were eligible for pensions, and for that working year Pontchartrain pensioned seventeen academicians. When Sédileau died in 1693, his pension was assigned to Varignon, and the number pensioned fell to sixteen. When Pothenot was excluded, probably in 1696, the number of pensioned members either fell again, to fifteen, or Boulduc was awarded Pothenot's pension, and the number pensioned remained sixteen. In 1698 Fontenelle began to receive his pension and the number of pensioned academicians rose to sixteen or seventeen. In 1699 twenty members of the Academy were designated *pensionnaires.* In addition, Truchet received a pension off the *estat* (table 2). In 1699, therefore, twenty-one academicians altogether were pensioned, twenty on the *estat* and one off it. From 1691 through 1699, therefore, Pontchartrain pensioned, on average, 16.36 members of the Academy of Sciences (fig. C.2).

Of the thirty-nine members of the Academy of Sciences from 1691 through 1698, probably nineteen, or 49 percent, were included on the *estat* (fig. 2.2).

Having calculated the amount spent on the Academy of Sciences and the number of academicians pensioned from 1691 through 1699, it is possible to estimate the size of the average pension (fig. C.2). From 1691 through 1699 Pontchartrain pensioned on average 16.36 members of the Academy of Sciences, and he budgeted for that purpose 269,550 livres or an average of 24,505 livres for each of the eleven working years. The average pension, therefore, would have been 1,498 livres. Cassini received 96,000 livres during the same period. Without him, the average annual cost of pensions was 15,777 livres, and the average pension was 1,027 livres.

By comparison, Pontchartrain pensioned all eight members of the Academy of Inscriptions, and he budgeted 144,767 livres from 1691 through 1699, or an average of 13,161 livres a year for that purpose. The average pension for members of the Academy of Inscriptions, therefore, was 1,645 livres. Tallemant, the best paid member, received 21,000 livres from 1691 through 1699. Without him, the average annual cost of pensions was 11,252 livres, and the average pension was 1,607 livres. Even the two translators received 1,200 livres apiece, which was more than the average pension paid to members of the Academy of Sciences other than Cassini. Of the three groups represented on the *estat* of the Academies, therefore, members of the Academy of Sciences received the lowest average pensions, when Cassini's exceptionally high pension is excluded from the calculations.

Pontchartrain's average pensions for members of the Academy of Sciences can also be compared with those of Colbert and Louvois before him (fig. C.2). From 1666 through 1683 Colbert pensioned on average 16.83 academicians and spent 614,525 livres for eighteen working years, or an average of 34,140 livres a year. The average pension, therefore, would have been 2,029 livres. But Huygens and Cassini received 216,500 livres of the total spent on pensions by Colbert. Without them, the average annual cost of pensions was only 22,113 livres; divided among, on average, 15.11 academicians, that gave an average pension of 1,463 livres. Louvois pensioned on average 15.86 academicians from 1684 through 1690 for seven working years, and he spent 159,500 livres, or an average of 22,786 livres a year. The average pension, therefore, would have been 1,437 livres. Huygens no longer received a pension, but Cassini's earnings under Louvois totaled 52,500 livres. Without Cassini, the average annual cost of pensions was only 15,286 livres; divided among, on average, 14.86 academicians, that gave an average pension of 1,029 livres. Of the three ministerial protectors, therefore, Colbert pensioned academicians the most generously, and Pontchartrain paid pensions at the lower levels established by Louvois.

Key to the Tables

Abbreviations

Abp.	Archbishop.
AN	Paris, Archives Nationales.
Anat.	*Anatomiste.*
ARdI	Académie royale des inscriptions.
ARdS	Académie royale des sciences.
Ast.	*Astronome.*
BdR	Bibliothèque du roi.
BN	Paris, Bibliothèque Nationale.
Bot.	*Botaniste.*
CAM	Compagnie des arts et métiers.
CdB	*Comptes des bâtiments.*
CdB	*Les comptes des bâtiments du roi sous le règne de Louis XIV,* ed. Guiffrey.
CdR	Cabinet du roi.
Chim.	*Chimiste.*
d.	Denier or deniers.
DA	*Dépense actuelle* of the Trésor royal.
F	Friday.
GC	*Grand comptant* of the Trésor royal.
Géom.	*Géomètre.*
Hist. an.	*Histoire des animaux.*
JR	Jardin royal.
lv.	Livre or livres.
M	Monday.
Méc.	*Mécanicien.*
Obs.	Observatoire.
PC	*Petit comptant* of the Trésor royal.
Pd.	Paid.
Phys.	*Physicien.*
PJ	*Petit jardin* of the JR.
Prés.	*Président.*
s.	Sol or sous.
Sa	Saturday.
Sec.	*Secrétaire.*
Th	Thursday.
Trés.	*Trésorier.*

Tu   Tuesday.
W   Wednesday.

Dates

In the tables, dates are given as follows: day. month. year, with the month represented by Roman numerals and years before 1700 by the last two digits of the year. In the notes, the names of the months are given in standard abbreviations.

The Categories of Research Expenditure, Tables 5–7

1   Shared Expenses: Physical Plant and Small Expenses.
    These include expenditures on behalf of the BdR and the JR, whose premises the Academy shared, or on behalf of several royal buildings, including the Obs., BdR, and JR. Note that this combines several categories treated separately in Stroup (*Company*, figs. 5.11 and 5.12).
2   Shared Expenses: Illustrations of Plants for CdR.
    Academicians had access to these drawings and engravings, which were executed by Robert and Joubert and kept in the BdR. Most records of expenditure in this category note that these illustrations continue a book of miniatures for the CdR at the BdR; the exceptions are in 1692 (AN G⁷), 1697 (*CdB*, 4: 187), and 1699 when Joubert was paid for four hundred drawings (*CdB*, 4: 477).
3   Shared Expenses: Rent.
    For the houses on rue Vivienne occupied by the BdR and ARdS.
4   Engravings and Drawings of Plants.
    Note that even when Chastillon's engravings of rare plants are said to be for the CdR, they have been included under ARdS expenses because that is consistent with the general pattern of expenditure. The illustrations of rare plants destined for the CdR were miniatures by Robert and Joubert, those for the ARdS were executed by Chastillon.
5   Scientific Instruments and Models of Machines.
    Note that this combines categories treated separately in Stroup (*Company*, figs. 5.2a, 5.2d, and 5.4a).
5* Shared Expenses: Scientific Instruments and Models of Machines.
    From 1695 the retainer paid to a clockmaker to maintain pendulum clocks no longer covered his services to the Academy exclusively, so that this item becomes a shared expense, as in categories 1, 2, 3, and 7*.
6   Chemical Laboratory.
7   *Petit jardin.*
7* Shared Expenses: *Petit jardin.*
    From 1693 expenditure on the PJ includes maintenance of the amphitheater and terrace, which were used by the JR, so that this becomes a shared expense, as in categories 1, 2, 3, and 5*.

8   Small Expenses.
9   Observatory.
10 Anatomical Research.
11 Engravings and Drawings of Animals.
12 Other Engravings and Drawings.
    The 3,097 lv. paid to Simonneau in 1699 was probably for engravings
    of *arts et métiers* in connection with the work of Des Billettes, Jaugeon,
    and Truchet.
13 Research on Minerals.
14 Research on Plants.

TABLE 1. The *Estat* of the Academies, 1689–1703

Fiscal Year	1689	
Working Year	1688	1688
Source	AN O¹ 1934ᴮ 14	*CdB*, 3: 305–307
Nature of Source	*Estat*	Record of Payment
Date of *Ordonnance*		
Amount Due on *Estat*	39,400 lv.	Pd. 8. I., 14. V. 90

Academicians, Assistants, & Translators	Amount Due lv.	Total lv.	Payments lv.	Total lv.
ARdS				
Cassini	9000		9000	
Borelly (d. 1689),* Homberg (1691–)	2000		2000	
Bourdelin	1500		1500	
Dodart	1500		1500	
Du Hamel	1500		1500	
Du Verneyᵇ	1500		1500	
Galloisᶜ	1500		1500	
La Hire	1500		1500	
Tournefort (1691–)	—		—	
Marchant	1200		1200	
Varignon	—		—	

TABLE 1. Continued

Academicians, Assistants, & Translators	Amount Due lv.	Total lv.	Payments lv.	Total lv.
Méry	600		600	
Couplet, C.-A.	500		500	
Sédileau (d. IV. 93)	500		500	
Pothenot	400		400	
Rolle	400		400	
Cusset	300		300	
Le Febvre	300		300	
Subtotal		24,200		24,200
Assistants				
Dalesme	600		600	
Du Verney's "garçon cirurgien"b	600		600	
Chastillon	400		400	
Subtotal		1,600		1,600
ARdI				
Boileau Despréaux	2000		2000	
Racine	2000		2000	
Charpentier	1500		1500	
La Chapelle (d. 1694), La Loubère (1694–)	1500		1500	
Rainssant (d. VI. 89), Tourreil (1691–)	1500		1500	
Renaudot (1691–)	—		—	
Tallemant	1500		1500	
Félibien (d. VI. 1695), Dacier (1695–)	1200		1200	
Subtotal		11,200		11,200

Translators		
De La Croix[d]	1200	
Dippy[d]	1200	
Subtotal		2,400
Total		39,400

Total Payments to Individuals	
Total Payment to Group	39,400
Total Paid	39,400

[a] The payments for 1688 went to Borelly and his widow.

[b] Du Verney was paid 2,100 lv., 1,500 for his pension and 600 for the wages of his assistant, a "garçon cirurgien" (BN MS. Clairambault 566: 251r), who is listed with the Academy's assistants.

[c] Although Gallois's pension under Louvois is ascribed to his work in *belles-lettres*, it was probably for his work in the Academy of Sciences, and his name was included with those of other members of the Academy of Sciences: BN MS. Clairambault 566: 247r, 251; AN G[7] 898 (June 1695) and 992; Gros de Boze, *Histoire de l'Académie des inscriptions*, historical introduction in vol. 1; Moréri, *Grand dictionnaire historique*, "Académie des inscriptions"; Maury, *Les Académies*; *L'ancienne Académie des inscriptions*, 6; Fabre, *Études littéraires*, 464–65, 467–68.

[d] De La Croix and Dippy were probably included in the *estat des gens de lettres* because they catalogued Turkish and Persian books and manuscripts for the Bibliothèque du roi: Franklin, *Anciennes bibliothèques*, 2: 190. For documents pertinent to them and their families, see AN G[7] 993 (14 Mar. 1695), 994 (24 May 1695), 995 (13 Mar. 1696), discussed in fiscal year 1694, note f; AN G[7] 893, 894, 898, 901, 902, for payments in addition to those summarized in tables 1 and 3; and AN PP 151.

TABLE 1. Continued

Fiscal Year	1690				1689			
Working Year	1689							
Source	AN O¹ 1934ᴮ 14				CdB, 3: 439–40			
Nature of Source	Estatᵃ				Record of Payment			
Date of Ordonnance								
Amount Due on Estat	35,900 lv.				Pd. 28. I. 90			
Academicians, Assistants, & Translators	Amount Due lv.	Total lv.	s.	d.	Payments lv.	s.	d.	Total lv.
ARdS								
Cassini	9000				3000			
Borelly (d. 1689), Homberg (1691–)	—				—			
Bourdelin	1500				500			
Dodart	1500				500			
Du Hamel	1500				500			
Du Verney	1500				500			
Gallois	1500				500			
La Hire	1500				500			
Tournefort (1691–)	—				—			

Marchant	1200	400
Varignon	—	—
Méry	600	200
Couplet, C.-A.	500	166. 13. 4
Sédileau (d. IV. 93)	500	166. 13. 4
Pothenot	400	133. 6. 8
Rolle	400	133. 6. 8
Cusset	300	100
Le Febvre	300	100
Subtotal	22,200	7,400
Assistants		
Dalesme	600	200
Du Verney's "garçon cirurgien"	600	200
Chastillon	400	133. 6. 8
Subtotal	1,600	533. 6. 8
ARdI		
Boileau Despréaux	2000	666. 13. 4
Racine	2000	666. 13. 4
Charpentier	1500	500
La Chapelle (d. 1694), La Loubère (1694–)	1500	500
Rainssant (d. VI. 89), Tourreil (1691–)	—	—
Renaudot (1691–)	—	—
Tallemant	1500	500
Félibien (d. VI. 1695), Dacier (1695–)	1200	400
Subtotal	9,700	3,233. 6. 8

TABLE 1. Continued

Academicians, Assistants, & Translators	Amount Due lv.	Total lv.	Payments lv.	s.	d.	Total lv.	s.	d.
Translators								
De La Croix	1200		400					
Dippy	1200		400					
Subtotal		2,400				800		
Total		35,900				11,966.	13.	4
Total Payments to Individuals								
Total Payment to Group								
Total Paid						11,966.	13.	4
Balance Owed						23,933.	6.	8

*The same *estat* was used for two consecutive years, 1688 and 1689, with the previous working year crossed off and other changes made to bring it up to date for 1689. At the end of the *estat* is the note "fonds fait en 1691 de 40000."

94

TABLE 1. Continued

Fiscal Year	1691	
Working Year	1689	
Source	BN MS. Clairambault 566: 247, 251	AN G⁷ 893
Nature of Source	Estat	GC, DA
Date of Ordonnance		22. IX. 91 (Sa)
Amount Due on Estat	23,933. 6. 8 lv.[a]	23,933. 6. 8 lv.
	Dated 7. VIII. 91	Pd. XI. 91

Academicians, Assistants, & Translators	Amount Due			Total			Payments			Total		
	lv.	s.	d.	lv.	s.	d.	lv.	s.	d.	lv.	s.	d.
ARdS												
Cassini	6000											
Homberg (1691–)[a]	—											
Bourdelin	1000											
Dodart	1000											
Du Hamel	1000											
Du Verney	1000											
Gallois	1000											
La Hire	1000											
Tournefort (1691–)[a]	—											
Marchant	800											
Varignon[b]	—											

95

TABLE 1. Continued

Academicians, Assistants, & Translators	Amount Due			Total			Payments			Total		
	lv.	s.	d.	lv.	s.	d.	lv.	s.	d.	lv.	s.	d.
Méry	400											
Couplet, C.-A.	333.	6.	8									
Sédileau (d. IV. 93)	333.	6.	8									
Pothenot[f]	266.	13.	4									
Rolle	266.	13.	4									
Cusset[f]	200											
Le Febvre	200											
Subtotal				14,800								
Assistants												
Dalesme	400											
Du Verney's "garçon cirurgien"	266.	13.	4									
Chastillon	400											
Subtotal				1,066.	13.	4						
ARdI												
Boileau Despréaux	1333.	6.	8									
Racine	1333.	6.	8									
Charpentier	1000											
La Chapelle (d. 1694), La Loubère (1694–)	1000											
Tourreil (1691–)[a]	—											
Renaudot (1691–)	—											
Tallemant	1000											
Félibien (d. VI. 1695–), Dacier (1695–)	800											
Subtotal				6,466.	13.	4						

Translators		
De La Croix	800	
Dippy	800	
Subtotal	1,600	
Total	23,933.	6.	8

Total Payments to Individuals			
Total Payment to Group	23,933.	6.	8
Total Paid	23,933.	6.	8

[a] Marginalia in the *estat* recommend a pension of 2,000 lv. for Tourreil and 600 lv. apiece for Homberg and Tournefort.
[b] According to BN MS. Clairambault 566: 251v, Varignon did not get a pension, although he had been a member for four years and deserved one.
[c] The author of BN MS. Clairambault 566 mistakenly copied the figures for Pothenot next to Cusset's entry; I have corrected this error by entering sums consistent with the *estat* in AN O[1] 1934[B]14 and the information in *CdB*, 3: 439–40.

TABLE 1. Continued

	1692			1691		
Fiscal Year						
Working Year	1690			1690		
Source	AN G⁷ 992		AN G⁷ 894	AN G⁷ 992		AN G⁷ 894
Nature of Source	Estat for 1690ᵃ		GC, DA	Estat for 1690		GC, DA
Date of Ordonnance			21. I. 92 (M)			22. IX. 92 (M)
Amount Due on Estat	35,900 lv.ᵇ		36,000ᵇ lv.	39,900 lv.		39,900 lv.
	Amount Due lv.	Total lv.	Payment lv.	Amount Due lv.	Total lv.	Payment lv.
Academicians, Assistants, & Translators	Dated I. 92		Pd. VII. 92	Including Marginalia		Pd. IV. 93
ARdS						
Cassini	9000			9000		
Homberg (1691–)	—			—		
Bourdelin	1500			1500		
Dodart	1500			1500		
Du Hamel	1500			1500		
Du Verney	1500			1500		
Gallois	1500			1500		
La Hire	1500			1500		
Tournefort (1691–)	—			—		

Marchant	1200	1200
Varignon[c]	300	300
Méry	600	600
Couplet, C.-A.	500	500
Sédileau (d. IV. 93)	500	500
Pothenot	400	400
Rolle	400	400
Cusset	—	—
Le Febvre[b]	300	300
Subtotal	22,200[b]	22,200
Assistants		
Dalesme	600	600
Du Verney's "garçon cirurgien"	600	600
Chastillon	400	400
Subtotal	1,600	1,600
ARdI		
Boileau Despréaux	2000	2000
Racine	2000	2000
Charpentier	1500	1500
La Chapelle (d. 1694), La Louibère (1694–)	1500	1500
Tourreil (1691–)[d]	—	[2000]
Renaudot (1691–)[d]	—	[2000]
Tallemant	1500	1500
Félibien (d. VI. 1695), Dacier (1695–)	1200	1200
Subtotal	9,700	13,700

99

TABLE 1. Continued

Academicians, Assistants, & Translators	Amount Due lv.	Total lv.	Payment lv.	Amount Due lv.	Total lv.	Payment lv.
Translators						
De La Croix	1200			1200		
Dippy	1200			1200		
Subtotal		2,400			2,400	
Total		35,900^b			39,900	
Total Payments to Individuals						
Total Payment to Group			36,000			39,900
Total Paid			36,000^b			39,900

[a] This *estat* was presented for approval in January 1692, but it bears none of the signatures denoting approval. It concludes with the formula, "Garde de mon tresor royal, M . . . , payez comptant . . . ," used by a secrétaire d'État: Mousnier, *Institutions*, 2: 194; *Correspondance des contrôleurs généraux*, 1: 578–79.

[b] Note the disparity between the amount due according to the *estat* and the amount paid by the treasury. The explanation may be that Le Febvre received 400 instead of 300 lv., the prose text states that he is entitled to 300 lv.; but the figures in the right-hand column of the *estat* show 400 ("iiij c") lv. See also appendix C.

[c] Varignon began to receive a pension when Cusset ceased to be paid.

[d] Gros de Boze, *Histoire de l'Académie des inscriptions*, 6–7, gives 1691 as the date of entry for Renaudot and Tourreil. This date is consistent with the archival evidence pertinent to pensions and has been adopted here, rather than that of 1690, given by Fabre, *Études littéraires*, 467–68. A marginal notation in AN G⁷ 992 points out that Renaudot and Tourreil have joined the Academy of Inscriptions and should be paid, but the payment of 36,000 lv. suggests that they were not included for 1690.

TABLE 1. Continued

	1694			1693		
Fiscal Year						
Working Year	1692			1693		
Source	AN G⁷ 897ᵃ			AN G⁷ 897ᵃ		
Nature of Source	GCᵇ			GCᵇ		
Date of *Ordonnance*	16. II. 94 (Tu)			28. IV. 94 (W)		
Amount Due on *Estat*	43,000 lv.			43,000 lv.		
	Pd. VI. 95–XII. 96			Pd. VI. 95–XII. 96		
Academicians, Assistants, & Translators	Payments lv.	Date of GC	Total lv.	Payments lv.	Date of GC	Total lv.
ARdS						
Cassiniᶜ	1500	VII. 95		9000	II–III. 96	
Homberg						
Bourdelin	1500	VII. 95		1500	VII. 95	
Dodartᵈ						
Du Hamel	1500	II–III. 96				
Du Verney	1500	VI. 95		1500	IV–VII. 96	
Gallois	1500	VI. 95		1500	VI. 95	
La Hire				1500	VI. 95	
Tournefort				1500	VIII. 96	
Marchant				1200	IV–VII.96	
Varignon						

TABLE 1. Continued

Academicians, Assistants, & Translators	Payments lv.	Date of GC	Total lv.	Payments lv.	Date of GC	Total lv.
Méry						
Couplet, C.-A.						
Sédileau (d. IV. 93)						
Pothenot						
Rolle						
Le Febvre						
Subtotal			7,500			17,700
Assistants						
Dalesme	600	XI. 95		600	XI. 95	
Du Verney's "garçon cirurgien"	600	II–III. 96		600	IV–VII. 96	
Chastillon						
Subtotal			1,200			1,200
ARdI						
Boileau Despréaux	2000	VI. 95		2000	II–III. 96	
Racine	2000	VI. 95		2000	II–III. 96	
Charpentier	1500	VI. 95		1500	XII. 95	
La Chapelle (d. 1694),° La Loubère (1694–)	1500	VII. 95		1500	IV–VI. 96	
Tourreil	2000	VI. 95		2000	VI. 95	
Renaudot	2000	VII. 95		2000	VII. 95	
Tallemant^f	1500	VI. 95		1500	VII. 95	
Félibien (d. VI. 1695), Dacier (1695)	1200	IX. 95				
Subtotal			13,700			12,500

Translators					
De La Croix[g]	1200	VI. 95	1200	VI. 95	
	1200	XI. 95	1200	XI. 95	
Dippy					
Subtotal		2,400		2,400	
Total					
Total Payments to Individuals	18,200	XII. 96	24,800	8,000	XII. 96	33,800
Total Payment to Group		18,200	18,200	8,000	8,000
Total Paid		43,000		41,800[a]	

[a] The record is not clear during this fiscal year, when the pensions for two working years were paid. The entries often do not refer to the *estat* of the person being paid. The entries sometimes do not specify which working year a pension represents, and occasionally give misleading information about the Academies, sometimes do not specify which working year a pension represents, and occasionally give misleading information about the person being paid. The most obscure entries are quoted in notes d, e, and f. One payment of 1,200 lv. for working year 1693 has not been identified, despite numerous searches; the missing item ought to be Félibien's pension. Thus, although the treasury records show the final payment of 8,000 lv. as completing payment of 43,000 lv., only payments totaling 41,800 lv. have been found. There is also an entry in Dec. 1694, "Au Sieur Marin, medecin de l'Accademie des sciences, par gratification . . . . 1200." This cannot be an item from the *estat* of the Academies, even if "Marin" is a misspelling of "Morin," because Morin de Toulon entered the Academy in Dec. 1693 and would not have been eligible for a pension before 1694, while the *estats* being paid in fiscal year 1694 were for 1692 and 1693.

[b] After Jan. 1694 the total of GC and PC does not equal that of DA, and the lump-sum payments of 18,200 lv. and 8,000 lv. do not appear in the DA. See appendix A.

[c] See appendix B, document V.

[d] The entries for July 1695 read "A Monsieur Dodart, docteur en medecine, pour ses appointemens 1692," and "A luy pour ses appointemens 1693." "Appointemens" was more commonly used for the salary Dodart received as physician to the king than for the stipend he earned as academician. Since Dodart received 3,000 lv. a year as physician and 1,500 lv. a year as academician, these entries have been taken as referring to his pension as academician, especially since they were paid at about the same time as pensions to Homberg, Gallois, and La Hire. For Dodart's stipend as physician, see Sept. 1694 and Feb. 1695.

[e] The date of La Chapelle's death is disputed, but it was probably Mar. 1694; his pensions for 1692 and 1693 were paid to his heirs.

[f] The entries for Tallemant's pensions are confused: "A Monsieur Paul Tallemant de l'Accademie françoise, par gratification 1694 suivant l'estat des professeurs . . . ," "1500" (June 1695) and "Au Sieur Paul Tallemant, intendant des devises et inscriptions des edifices royaux, par gratification 1693 suivant l'estat . . . ," (July 1695).

[g] The records of De La Croix's pensions for 1692 through 1694 are inconsistent. AN G[7] 897, June 1695, GC, lists two payments to him: "Au Sieur de la Croix Petit, secretaire interprette du roy, en langue turc, pour ses appointemens 1692, suivant l'estat 1694 . . . . 1200," and "A luy, pareille somme, pour ses appointemens 1693 en lasdite qualite suivant l'estat . . . . 1200." But AN G[7] 994 (24 May 1695) contains a petition from De La Croix, "employé sur l'estat de l'Académie des Sciences," requesting conversion of his pensions from 1692, 1693, and 1694 into *rentes*; this was approved "au denier 18 en doublant" for 1,800 lv. instead of 3,600 lv.

TABLE 1. Continued

Fiscal Year	1695	
Working Year	1694	
Source	AN G⁷ 898	
Nature of Source	GCª	
Date of *Ordonnance*	22. I. 95 (F)	
Amount Due on *Estat*	43,000 lv.	

Pd. VI. 95–XII. 96

Academicians, Assistants, & Translators	Payments lv.	Date of GC	Total lv.
ARdS			
Cassini			
Homberg	1500	IX. 96	
Bourdelin			
Dodart	1500	VII. 95	
Du Hamel			
Du Verney	1500	VI. 96	
Gallois	1500	VI. 95	
La Hire	1500	VI. 95	

Tournefort			
Marchant			
Varignon[b]			
Méry			
Couplet, C.-A.			
Pothenot			
Rolle			
Le Febvre			
Subtotal			7,500
Assistants			
Dalesme	600	I. 96	
Du Verney's "garçon cirurgien"	600	VI. 96	
Chastillon	400	X. 96	
Subtotal			1,600
ARdI			
Boileau Despréaux			
Racine			
Charpentier			
La Chapelle (d. 1694), La Loubère (1694–)[c]			
Tourreil	2000	VI. 95	
Renaudot			
Tallemant	1500	VII. 95	
Félibien (d. VI. 1695),[d] Dacier (1695)	1200	V. 96	
Subtotal			4,700

105

TABLE 1. Continued

Academicians, Assistants, & Translators	Payments lv.	Date of GC	Total lv.
Translators			
De La Croix	1200	VI. 95	
Dippy[e]	1200	VII. 96	
Subtotal			2,400
Total			
Total Payments to Individuals			16,200
Total Payment to Group	26,800	XII. 96	26,800
Total Paid			43,000

[a] From Sept. 1696 through July 1697 the total of GC and PC does not agree with that of DA. The lump-sum payment of 26,000 lv. listed in GC for Dec. 1696 was not recorded in DA. See appendix A.

[b] Varignon was awarded the amount of the deceased Sédileau's pension.

[c] Although Racine (Oeuvres, 7: 78, n. 19) indicates that La Loubère, in favor with Pontchartrain, became an academician in 1693, he would not have received a pension before La Chapelle died.

[d] The petition of Félibien's widow, Marguerite Le Maire, for the pensions due him for 1694 and 1695 survives in AN G^7 996, 21 Mar. 1696.

[e] A petition from Pierre Dippy in AN G^7 995 (13 Mar. 1696) describes him as "arabe de nation professeur au Collège royal de France en la langue arabesque" and as having served the king for forty-four years. He and his family depended for their subsistence on the pension he received from the king, but he had not received his 1694 or 1695 payments. Pontchartrain authorized payment for 1694.

TABLE 1. Continued

	AN G⁷ 899			AN G⁷ 900		AN G⁷ 899–900
Fiscal Year	1696					
Working Year	1695					
Nature of Source	GCᵃ			GCᵇ		
Date of *Ordonnance*	?					
Amount Due on *Estat*	44,000 lv.					
	Pd. III. 96–II. 97			Pd. between XII. 97 & I. 99		
Academicians, Assistants, & Translators	Payments lv.	Date of GC	Total lv.	Payments lv.	Date of GC	Total lv.
ARdS						
Cassini	1500	VI. 96				
Homberg						
Bourdelin						
Dodart						
Du Hamel						
Du Verney						
Gallois						
La Hire						
Tournefort						
Marchant						

107

TABLE 1. Continued

Academicians, Assistants, & Translators	Payments lv.	Date of GC	Total lv.	Payments lv.	Date of GC	Total lv.
Varignon						
Méry						
Couplet, C.-A.						
Pothenot						
Rolle						
Le Febvre						
Subtotal			1,500			
Assistants						
Dalesme						
Du Verney's "garçon cirurgien"	600	VI. 96				
Chastillon						
Subtotal			600			
ARdI						
Boileau Despréaux						
Racine						
Charpentier						
La Loubère						
Tourreil	2000	VII. 96				
Renaudot						
Tallemant	2500	V. 96, II. 97				
Félibien (d. VI. 95),c Dacier (1695–)	550	VI. 96				
Subtotal			5,050			

108

Translators					
De La Croix	1200	III. 96			
Dippy[d]					
Subtotal		1,200			
Total			8,350		8,350
				35,650[e]	28. XII. 97–26. I. 99	35,650
Total Payments to Individuals			8,350		
Total Payment to Group					
Total Paid			44,000		

[a] In AN G⁷ 899, the total of GC and PC does not equal that of DA during the months when academicians' pensions are listed in GC. See appendix A.

[b] In AN G⁷ 900, there is no DA corresponding to the GC in which the lump-sum payment to academicians was listed, and the accounts for the end of the fiscal year do not balance.

[c] See fiscal year 1695, note d.

[d] See fiscal year 1695, note e.

[e] This payment is recorded in a document headed: "M. Brunet 1696 26 janvier 1699. Depense du Tresor Royal depuis le 28 decembre 1697." One item reads: "Aux academies reste de 44000 pour 1695 . . . . 35650." The document is in the folder entitled "Monsieur de Turmenyes 1696. Feuilles de la recepte et de la depense depuis le premier janvier 1699. Pour les veriffier et y ajouster cequi a esté reçu depuis jusqu'à ce jourdhuy 22 juin 1701." In the folder are three smaller folders. The entry concerning the Academy is in that headed "Memoire géneral sur l'estat au vray 1696." Two other references to the *estat* appear in these papers, under the heading of "gratiffications." The late date of this payment recalls père Léonard's gossip about the 1695 pensions; see chap. 1, n. 13.

TABLE 1. Continued

	1697			1697		
Fiscal Year						
Working Year	1696			1697		
Source	AN G⁷ 901			AN G⁷ 901		
Nature of Source	GCᵉ			GCᵉ		
Date of Ordonnance	14. II. 97 (Th)			31. XII. 97 (Tu)		
Amount Due on Estat	44,000 lv.			44,000 lv.		
	Pd. I. 98			Pd. XII. 98		
	Payments lv.	Date of GC	Total lv.	Payments lv.	Date of GC	Total lv.
Academicians, Assistants, & Translators						

ARdS
  Cassini
  Homberg
  Bourdelin
  Dodart
  Du Hamel
  Du Verney
  Gallois
  La Hire
  Tournefort

Marchant
Varignon
Méry
Couplet, C.-A.
Boulduc?
[Pothenot stopped attending in 1696]
Rolle
Le Febvre

Subtotal

Assistants
Dalesme
Du Verney's "garçon cirurgien"
Chastillon

Subtotal

ARdl
Boileau Despréaux
Racine
Charpentier
La Loubère
Tourreil
Renaudot
Tallemant
Dacier

Subtotal   2500   IX. 98   2,500

TABLE 1. Continued

Academicians, Assistants, & Translators	Payments lv.	Date of GC	Total lv.	Payments lv.	Date of GC	Total lv.
Translators						
De La Croix						
Dippy	1200	XI. 97		1200	XI. 98	
Subtotal			1,200			1,200
Total						
Total Payments to Individuals	42,800	I.98	1,200	40,300	XII. 98	3,700
Total Payment to Group	42,800		42,800	40,300		40,300
Total Paid			44,000			44,000

* There is no monthly DA for payments in 1698, but the totals of both *estats* are included in the summary of "Despenses par chapitre."

TABLE 1. Continued

Fiscal Year:	1698	
Working Year	1698	
Source	AN G[7] 902	
Nature of Source	GC[a]	
Date of Ordonnance	31. XII. 98 (W)	
Amount Due on Estat	45,500 lv.	
	Pd. III. 99–III. 1700	
Academicians, Assistants, & Translators	Payments lv.	Date of GC	Total lv.

ARdS
Cassini
Homberg
Bourdelin
Dodart
Du Hamel
Du Verney
Gallois
La Hire
Tournefort

TABLE 1. Continued

Academicians, Assistants, & Translators	Payments lv.	Date of GC	Total lv.
Fontenelle ?			
Marchant			
Varignon			
Méry			
Couplet, C.-A.			
[Pothenot]			
Rolle			
Boulduc ?			
Le Febvre			
Subtotal			
Assistants			
Dalesme			
Du Verney's "garçon cirurgien"			
Chastillon			
Subtotal			

114

ARdI			
Boileau Despréaux			
Racine			
Charpentier	1500	III. 99	
La Loubère			
Tourreil			
Renaudot			
Tallemant			
Dacier			
Subtotal			1,500
Translators			
De La Croix			
Dippy			
Subtotal			
Total			
Total Payments to Individuals	44,000	III. 1700	1,500
Total Payment to Group			44,000
Total Paid			45,500

* No DA survives for 1698, but the sum of the *estat* is included in the summary of "Despenses par chapitre" and in the large worksheet used as a folder.

TABLE 1. Continued

Fiscal Year	1699
Working Year	1699
Source	AN G⁷ 903
Nature of Source	"Despenses 1699 a arrester"[a]
Date of *Ordonnance*	26. I. 1700 (Tu)
Amount Due on *Estat*	47,800 lv.[b]

Academicians, Assistants, & Translators	Hypothetical *estat*[c]		Pd. by III. 1700		
	Amount Due lv.	Total lv.	Payment lv.	Date	Total lv.
ARdS					
Cassini	9000				
Bourdelin	1500				
Dodart	1500				
Du Hamel	1500				
Du Verney	1500				
Fontenelle	1500				
Gallois	1500				
Homberg	1500				
La Hire	1500				
Tournefort	1500				

Marchant	1200
Des Billettes	1000
Jaugeon	1000
Méry	600
Rolle	400
Varignon	800
Couplet, C.-A.	500
Dalesme	600
Boulduc	600
Lémery[d]	400
Le Febvre	400
Subtotal	29,400
Assistants	
Chastillon	400
Du Verney's "garçon cirurgien"	600
Subtotal	1,000
ARdI	
Boileau Despréaux	2000
Racine (d. IV. 99), Pavillon	2000
Charpentier	1500
La Loubère	1500
Tourreil	2000
Renaudot	2000
Tallemant	2500
Dacier	1500
Subtotal	15,000

TABLE 1. Continued

Academicians, Assistants, & Translators	Amount Due lv.	Total lv.	Payment lv.	Date	Total lv.
Translators					
De La Croix	1200				
Dippy	1200				
Subtotal		2,400			
Total		47,800			
Total Payments to Individuals					
Total Payment to Group			47,800	by III. 1700	47,800
Total Paid					47,800[e]

[a] See discussion of AN G⁷ 903 in appendix A.

[b] [Chamillart], "État," includes this item in the accounts payable on 3 Oct. 1699.

[c] This hypothetical *estat* is based on the list of *pensionnaires* of the Academy of Sciences, as established at the time of the *règlement* of 1699, and the *estat* of 1703, in which Le Febvre was replaced by Maraldi and Bourdelin by Lémery. See appendix C.

[d] Lémery was appointed *pensionnaire* 28 Nov. 1699, probably too late to receive a pension that year.

[e] In a document dated 3 Mar. 1700 which summarizes payments due, the total of the *estat* for the Academies appears as 46,300, not 47,800, lv., a discrepancy of 1,500 lv. (AN G⁷ 973). Perhaps one academician had already been paid, or perhaps the lower sum represents a deduction from the *estat* of the remaining three-quarters of Racine's pension after he died in April 1699. But the records show that the sum of 47,800 lv. was paid, perhaps because Pavillon was appointed to replace Racine and received a pension of 1,500 lv.

TABLE 1. Continued

Fiscal Year	1700	1701	1703
Working Year	1700	1701	1703
Source	AN G⁷ 973		
Nature of Source	Payments Due	Payments Due	Estat
Date of Ordonnance			?
Amount Due	47,800	49,400[a]	52,400
Date of Document	?	2. IX. 1702	?

Academicians, Assistants, & Translators, 1703	Amount Due lv.	Amount Due lv.	"Gratifications" lv.	"Augmentation" lv.	Subtotal lv.	Total Due lv.
ARdS						
Cassini			9000			
Dodart			1500			
Du Hamel			1500			
Du Verney			1500			
Fontenelle			1500			
Gallois			1500			
Homberg			1500			
La Hire			1500			
Tournefort			1500			

119

TABLE 1. Continued

Academicians, Assistants, & Translators, 1703	Amount Due lv.	"Gratifications" lv.	"Augmentation" lv.	Sub-Total lv.	Total Due lv.
Marchant		1200			
Des Billettes		1000			
Jaugeon		1000			
Méry		900			
Rolle		900			
Varignon		900			
Couplet, C.-A.		600			
Dalesme		600			
Boulduc		500			
Lémery		400			
Maraldi		400			
Subtotal					29,400
Assistants					
Chastillon		600	400		
Du Verney's "garçon cirurgien"					
Subtotal					1,000

120

ARdI			
Tallemant	1500	2000	3500
Tourreil	2000	1000	3000
Dacier	1500	1000	2500
Boileau Despréaux	2000	—	2000
Renaudot	2000	—	2000
Pavillon	1000	1000	2000
La Loubère	1000	500	1500
Charpentier (d. 1702), Vaillant (1702–)	1000	100	1100
Boutard	1000	—	1000
Félibien	1000	—	1000
Subtotal			19,600
Translators			
Dippy	1200		
Petit de la Croix	1200		
			2,400
Total	47,800	49,400	52,400

*The increase of 1,600 lv. in 1701 must be on behalf of the Academy of Inscriptions, which received its *règlement* this year. The increase in the *estat* for 1703 represents *augmentations* for members of the Academy of Inscriptions.

TABLE 2. Pensions Paid to the Academy of Sciences under Pontchartrain, 1691–99

Fiscal Year	Working Year	Paid on the Estat[a]			Paid off the Estat[b]		Total Paid On & Off the Estat	Fiscal Year
		Academicians	Assistants	Subtotal	Couplet[c]	Truchet[d]		
1691	1689	14,800	1,066. 13. 4	15,866. 13. 4			15,866. 13. 4	1691
1692	1690	22,300	1,600	23,900			23,900	1692
1692	1691	22,200	1,600	23,800			23,800	1692
1694	1692	25,300	1,600	26,900			26,900	1694
1694	1693	25,300	1,600	26,900			26,900	1694
1695	1694	25,300	1,600	26,900			26,900	1695
1696	1695	25,300	1,600	26,900			26,900	1696
1697	1696	25,300	1,600	26,900	300		27,200	1697
1697	1697	25,300	1,600	26,900			26,900	1697
1698	1698	26,800	1,600	28,400	500		28,900	1698
1699	1698							
1699	1699	29,400	1,000	30,400	450	1000	31,850	1699
Total		267,300	16,466. 13. 4	283,766. 13. 4	1,250	1000	286,016. 13. 4	

a See fig. C.1, appendix C, and table 1, fiscal year 1692, note b.

b Charas and Morin also received payments off the estat. Charas as a new convert (table 3) and Morin for reasons which were not specified. The entries for Morin are not necessarily for work done in the Academy. The entry in AN G^7 897 is quoted in table 1, fiscal year 1694, note a. AN G^7 898, Dec. 1695 (PC), 899, Dec. 1696 (PC), and 902, Dec. 1698 (PC), list pensions of 600 lv. for "Morin medecin." These payments to Charas and Morin have not been included in the calculations of pensions paid to academicians.

c These payments are to Couplet as concierge to the Observatory: CdB, 4: 268, 411, 427, 566. From 1683 through 1695 C.-A. Couplet had earned his pension on the estat as "concierge de l'Observatoire et commis à la garde et entretien des instruments et machines." The year he was appointed treasurer, 1696, payments to Couplet as "concierge de l'Observatoire" began to appear in the CdB. Either C.-A. Couplet accumulated two pensions in connection with the Academy, or his son P. Couplet was paid off the estat for assuming the responsibilities of concierge. Given the coincidence of these changes in 1696, it is plausible that C.-A. Couplet relinquished his post as concierge, since no position in the Academy was a sinecure. Furthermore, neither the Academy's records nor Wolf's Observatoire mention P. Couplet's assumption of this post in the 1690s. It has been assumed, therefore, that C.-A. Couplet remained concierge and collected two incomes for his work at the Academy. The payments to Couplet as concierge to the Observatory continue in the eighteenth century at the rate of 500 lv. a year: CdB, 4: 683, 689, 805, 921, 1028, 1134, 1243.

d Like Des Billettes and Jaugeon, after Truchet joined the Academy of Sciences, he continued to collect the pension of 1,000 lv. he had received for his work in the Compagnie des arts et métiers. Since the règlement of the Academy forbade pensioning members of religious orders, Truchet received his pension off the estat: AN G^7 986–87, 26 Apr. 1699 (for 1699), and 25 Apr. 1701 (for 1700).

TABLE 3. Pensions Paid by the Royal Treasury, 1690–98

Amount Paid	Académie des Sciences[1]	Académie des Inscriptions[2]	Collège Royal[3]	Jardin Royal[4]	Other[5]
9,000 lv. and more	Cassini				Fagon, Premier médecin du roi (897) Du Chesne, Premier médecin des princes (897)
6,000 lv.				Dacquin & Fagon, Surintendants	Dacquin, Premier médecin du roi (897) Dionis, Chirurgien de feue Mme la Dauphine (897)
3,000 lv.				Dacquin, Surintendant	Renau, Ingénieur (892) Thévenot, Bibliothécaire du roi (893)
2,500 lv.		Tallemant			
2,400 lv.					Bourdelot, Médecin ordinaire du roi (897)
2,000 lv.		Boileau Despréaux Racine Renaudot Tourreil	De Launay & Germain, Droit français (892, 902)	Brément, Jardinier	Mlle de Scudéry (893) Seron, Médecin des bâtiments (CdB, 3: 639) Tarade, Ingénieur (894)
1,700–1,800 lv.			Gallois, Langue grecque (895, 898)		
1,500 lv.	Bourdelin Dodart Du Hamel Du Verney Gallois Homberg La Hire Tournefort	Charpentier Dacier La Chapelle Tallemant		Dacquin le jeune, Du Verney, & Fagon, Démonstrateurs	Boivin, BdR (897) Bose, Interprète d'anglais (897) Charas, Cy-devant de la R.P.R. (899) Niquet, Ingénieur (892)

123

TABLE 3. Continued

Amount Paid	Académie des Sciences[1]	Académie des Inscriptions[2]	Collège Royal[3]	Jardin Royal[4]	Other[5]
1,200 lv.	Marchant	Dacier, Félibien	Boudouin, Doyen (897); Gerbais, Langue et éloquence latine (994); Noel, Philosophie grecque (895)	Fagon, Sous-démonstrateur	Widow La Quintinie. La Quintinie had been gardener to the king (892); Chazerat, Ingénieur (892); Dippy & De La Croix, Interprètes (table 1)
1,050–1,100 lv.			Dippy, Langue arabe (897, 898, 901)		
1,000 lv.			Baleuze, Droit canon (897); Le Cappelain, Langue arabe (897)		De Vieussen, Médecin à Montpellier (892); Pernet, Ingénieur (892); Des Billettes, Jaugeon, & Truchet, CAM (chap. 6); Obry, Cy-devant de creance juifve (894, 897, 899)
900 lv.			Enguehard, Médecine & Pharmacie (994); La Hire, Mathématiques (898)		Robert, Prof. royal de Théologie (895)
800 lv.	Varignon		La Hire, Mathématiques (895)		
750 lv.			Langlet, Éloquence latine (895)		
700 lv.			Sauveur, Mathématiques (895)		

124

600–690 lv.	Dalesme Du Verney's aide Méry	Petit De La Croix, Langue arabe (897, 994) Enguehard, Médecine (895, 897, 994) Germain Preaux, Médecine (895)		Dubose, Ingénieur (892) Paul Hosti, S. J., Directeur des aumôniers de la marine, Prof. en mathématiques (902) Joly, Médecin à Montpellier (892) Papin, Cy-devant ministre de la R.P.R. à Hambourg (902) Seron, Médecin (902)
500 lv.	Couplet Sédileau	Petit De La Croix, Langue arabe (895, 994) Le Marie, Droit canon (895)	Boulduc	De La Vigne, Ingénieur (892) Denis, Maître de mathématiques des pages de Monseigneur (892)
450 lv.			Chaillou, Portier	
400 lv.	Chastillon Le Febvre Pothenot Rolle			Le Peintre, Joueur de violon (897)
300 lv.	Le Febvre Varignon			Bouchard, Ingénieur (893)
200 lv.	Danjau, Médecine (895)	Beaupré & Guarigues, Garçons du laboratoire		Collot, Ingénieur (895)

[1] See table 1.

[2] See table 1.

[3] References in parentheses are to AN G⁷. The *gages et augmentations* of four professors royal were increased: Gallois, from 1,700 to 1,800 lv.; La Hire, from 800 to 900 lv.; Enguehard, from 623 lv. 6 s. to 690 lv., and then to 900 lv.; and De La Croix, from 500 to 600 lv.

[4] *CdB*, 3: 438, 581, 729–31, 863, 997, 1011–12, 1125, 1143, 1145, 1152; 4: 54–55, 72, 198, 210, 217, 345, 362, 488. Dacquin received 3,000 lv. as *surintendant* in 1690, but from 1691 through 1694 he received 6,000 lv., the additional sum being paid as *augmentations*. See appendix B, document VI, for an explanation of this practice. In November 1694 Fagon replaced Dacquin as *surintendant* and kept his appointments as *démonstrateur* and *sous-démonstrateur*. Through 1694 the payments to Boulduc are designated as reimbursement for expenses and payment for teaching of a course in chemistry. After 1694 only reimbursements are mentioned.

[5] Unless otherwise noted, references in parentheses are to AN G⁷. Charas and Papin received payment as converts from Protestantism (la Religion prétendue réformée, or R. P. R.), as did Obry, as a convert from Judaism. Is this the same Obry who supplied apparatus to the Academy?

TABLE 4. Absences from Meetings of the Academy of Sciences, 1695–98

Academicians Eligible to Attend	Date of Admission	Date of Death or Exclusion	Title	Absences from 82 Meetings at which Botany was Discussed					
				1695 M/E	1696 M/E	1697 M/E	1698–99 M/E	Total M/E	% Absences
Du Hamel[P]	1666	6. VIII. 1706	Sec., Anat.	0/18	0/21	3/21	0/22	3/82	3.7
Bourdelin[P]	1666	14. X. 99	Chim.	0/18	0/21	1/21	0/22	1/82	1.2
Couplet, C.-A.[P]	1666	25. VII. 1722	Trés., Méc.	9/18	4/21	9/21	15/22	37/82	45.1
Gallois[P]	1668	19. IV. 1707	Géom.	5/18	5/21	2/21	2/22	14/82	17.1
Cassini I[P]	1669	14. IX. 1712	Ast.	18/18	5/21	0/21	0/22	23/82	28
Dodart[P]	1671	5. XI. 1707	Bot.	5/18	8/21	8/21	3/22	24/82	29.3
Du Verney[P]	1674	10. IX. 1730	Anat.	11/18	19/21	16/21	21/22	67/82	81.7
Leibniz*	1675	14. XI. 1716	Hon.						100
La Hire, P. de[P]	1678	21. IV. 1718	Ast.	0/18	5/21	2/21	2/22	9/82	11
Marchant, J.[P]	1678	11. XI. 1738	Bot.	0/18	2/21	1/21	4/22	7/82	8.5
Tschirnhaus*	1682	11. IX. 1708	Géom.						100
Pothenot[P]	1682	Ex. by 1699, or, if in 1696	Géom.	7/18 7/18	21/21 0/0	21/21 0/0	22/22 0/0	71/82 0/0	86.6 38.9
Le Febvre[P]	1682	1706	Ast.	7/18	8/21	10/21	20/22	45/82	54.9
Méry[P]	1684	3. XI. 1722	Anat.	0/18	1/21	3/21	1/22	5/82	6.1
Rolle[P]	1685	8. XI. 1719	Géom.	6/18	5/21	14/21	10/22	35/82	42.7
Varignon[P]	1688	23. XII. 1722	Géom.	2/18	2/21	3/21	1/22	8/82	9.8
Bignon*	1691	14. III. 1743	Prés.						100
Tournefort[P]	21. XI. 91	28. XII. 1708	Bot.	9/18	16/21	10/21	12/22	47/82	57.3

	Date								%
Homberg[P]	24. XI. 91	24. IX. 1715	Chim.	3/18	6/21	4/21	3/22	16/82	19.5
Charas	19. IV. 92	21. I. 98	Chim.	1/18	9/21	21/21	0/0	31/60	51.7
La Coudraye	10. VI. 93	unknown	Géom.						100
L'Hospital*	17. IV. 93	3. II. 1704	Géom.	17/18	21/21	21/21	22/22	81/82	98.8
Morin de Toulon	16. XII. 93	1707	Bot.	6/18	9/21	10/21	19/22	44/82	53.7
Cassini II†	12. VI. 94	15. IV. 1756	Ast.	18/18	5/21	5/21	1/22	29/82	35.4
La Hire, G.-P.†	19. VI. 94	4. VI. 1719	Ast.	2/18	9/21	6/21	2/22	19/82	23.2
Boulduc[P]	7. VIII. 94	23. II. 1729	Chim.	9/18	10/21	8/21	14/22	41/82	50
Maraldi†	28. VIII. 94	1. XII. 1729	Ast.	3/18	2/21	3/21	1/22	9/82	11
Chazelles*	1695	16. I. 1710	Ast.						100
Lagny*	11. I. 96	12. IV. 1734	Géom.	0/0	19/21	18/21	22/22	59/64	92.2
Sauveur	29. II. 96	9. VII. 1716	Géom.	0/0	11/18	20/21	22/22	53/61	86.9
Couplet, P.†	4. IV. 96	23. XII. 1743	Méc.	0/0	3/17	10/21	22/22	35/60	58.3
Guglielmini*	1696	11. VII. 1710	Phys.						100
Fontenelle[P]	9. I. 97	9. I. 1757	Sec., Géom.	0/0	0/0	7/20	2/22	9/42	21.4
Carré†	16. III. 97	11. IV. 1711	Géom.	0/0	0/0	3/16	0/22	3/38	7.9
Tauvry†	30. IV. 98	7. II. 1701	Anat.	0/0	0/0	0/0	1/20	1/20	5
Langlade*	19. XI. 98	VI. 1717	Chim.	0/0	0/0	0/0	1/20	1/20	100

Key

†   Student member.

*   Associate, external, or honorary member.

Ex.   Excluded. Pothenot was the only academician excluded during the 1690s; see appendix C for the possibility that this occurred in 1696.

M/E   Number of meetings Missed/Number of meetings which the member was Eligible to attend.

P   Pensioned. Note assumption that Pothenot was pensioned through 1695, Boulduc from 1696 (see appendix C and fig. 2.2).

1698–99   Plants were discussed at only one meeting in 1699 before the *règlement*.

Sources: AdS, Reg., 14–18 (attendance), CdB, IB (dates and titles).

TABLE 5. Research Expenses of the

Fiscal Year 1690		To Whom Paid	Amount Paid					
			ARdS Expenses (Categories 4–14)			Shared Expenses (Categories 1–3)		
Working Year	Purpose of Expenditure		livres	sous	deniers	livres	sous	deniers
1687	Expenses of BdR	Thévenot				1,998	0	6
1688	Carpentry at BdR & JR	Pierre Guerin, carpenter				1,300	0	0
1687–89	Laboratory	Bourdelin	1,019	4	6			
	Repairs to astronomical instruments	Gosselin, armorer	100	0	0			
1688–89	Glazing at BdR	Charles-François Jacquet, glazier				180	18	5
1688–90	Expenses of continuing *Hist. an.*	Du Hamel	316	10	0			
1689	Maintenance of PJ	Marchant	89	0	0			
	Rent of 2 buildings occupied by BdR	Abp. of Rouen				5,000	0	0
1689–90	Small expenses	Couplet	65	17	6			
	Salary of porter at Obs.	Seignelonge	150	0	0			
1690	Livery for new porter at Obs.	Antoine Voirie	30	0	0			
	Masonry at Obs.	Le Pas, contractor	300	0	0			
	Repairs to Obs.	La veuve Janson, glazier				600	0	0
	12 drawings of rare plants, on vellum in continuation of book of miniatures	Jean Joubert, painter				600	0	0
	9 pages of drawings of rare plants for the natural history of plants	Louis Chastillon, draftsman & engraver	198	0	0			
	6 copper plates etched with aqua fortis, showing rare plants, for the cabinet of engravings, BdR	Chastillon	528	0	0			
	Retainer for maintenance of mathematical instruments	Gosselin & Lagny, armorers	200	0	0			
	Retainer for maintenance of all pendulum clocks at AdS & Obs.	Isaac Thuret, clockmaker	300	0	0			
	Total Fiscal Year 1690		3,296	12	0	9,678	18	11

* Information is organized by the fiscal year to which payment was charged and, within fiscal year, according to the working year in which the obligation was incurred.

Financial obligations incurred on behalf of the Academy were often met a year or more later. Thus, some payments charged to fiscal year 1700 or later should also shed light on the Academy's activities during the 1690s. The *CdB*, for example, lists nine payments in fiscal year 1700 for work done previously. One (4: 625) was for maintenance of the PJ in 1698 and 1699, and eight (4: 616–

Source					Total		
CdB	AN G⁷	BN MS.	Notes	Category	livres	sous	deniers
3:438–39				1			
3:427				1			
3:428				6			
3:441				5			
3:429				1			
3:441				10			
3:438				7			
3:442				3			
3:441				8			
3:441				9			
3:441				9			
3:425				9			
3:428–29				1			
3:430				2			
3:433				4			
3:433				4			
3:503				5			
3:503				5			
					12,975	10	11

17, 621–22, 630, 689–90, 695) were for rent, maintenance of buildings and clocks, porter and concierge at the Obs., and drawings of rare plants in 1699. But these expenditures indicate little more than continuity at minimal levels. Expenditure during the early 1700s for the Academy's earlier work has not been sought in AN G⁷ 905 and subsequent boxes, because these records provide even fewer details than do those for the previous decade.

TABLE 5.

| Fiscal Year 1691 | | | Amount Paid | | | | | |
| | | | ARdS Expenses (Categories 4-14) | | | Shared Expenses (Categories 1-3) | | |
Working Year	Purpose of Expenditure	To Whom Paid	livres	sous	deniers	livres	sous	deniers
1689	Woodwork at BdR	Pierre Guérin				300	0	0
	Expenses of BdR	Thévenot				902	14	0
	Expenses of BdR	Thévenot				156	15	6
1689–90	Maintenance at BdR	Jean Gombault, glazier				200	0	0
1690	Laboratory	Bourdelin	240	7	0			
	Maintenance of PJ	Marchant	63	4	0			
	Rent of 2 buildings occupied by BdR	Abp. of Rouen				5,000	0	0
	Small expenses for maintenance of sites & for exercises and experiments of AdS	Couplet	180	9	0			
	Woodwork & repairs at Obs.	Gabriel Rozier, carpenter	96	10	0			
1690–91	Porter at Obs., 21 months' salary, plus livery	Voirie	180	0	0			
			142	10	0			
	36 drawings of plants, continuation of book of miniatures for CdR	Joubert				600	0	0
						300	0	0
	Metalwork at Obs.	Jean Blancheton, locksmith				150	0	0
						140	0	0
	Repairs & maintenance at Obs.	La veuve Janson				600	0	0
	6 copper plates of rare plants for cabinet of engravings, BdR	Chastillon	528	0	0			
1691	Retainer for mathematical instruments	Gosselin & Lagny	200	0	0			
	Retainer for pendulum clocks	Thuret	300	0	0			
	Subtotals							
	Paid during Louvois's protectorship to July '91		1,288	10	0	7,752	14	0
	Paid during Pontchartrain's protectorship from July '91		642	10	0	596	15	6
	Total Fiscal Year 1691		1,931	0	0	8,349	9	6

[a] Pd. 7 Jan. 1691.
[b] Pd. 7 Jan., 8 Apr. 1691.
[c] Pd. 7 Aug. 1691.
[d] Pd. 21 Jan. 1691.
[e] Pd. 29 Apr. 1691.
[f] Pd. 6 May 1691.
[g] Pd. 3 June 1691.
[h] Pd. 11 Mar. 1691.
[i] Pd. 10 June 1691.
[j] Pd. 8 Apr. 1691. The porter received 150 lv. as salary and 30 lv. for livery each year.

[k] Pd. 21 Oct. 1691, 13 Jan. 1692.
[l] Pd. 7 Jan., 8 Apr. 1691.
[m] Pd. 4 Nov. 1691.
[n] Pd. 11 Mar., 17 Apr. 1691.
[o] Pd. 4 Nov. 1691.
[p] Pd. 7 Jan., 18 Mar. 1691.
[q] Pd. 7 Jan. 1691.
[r] Pd. Jan. 1692?
[s] Pd. Jan. 1692?

	Source				Total		
CdB	AN O¹	BN MS.	Notes	Category	livres	sous	deniers
3:567			a	1			
3:582	2173:150r		b	1			
3:582	2173:150r		c	1			
3:569			d	1			
3:584		n. a. fr. 5147	e	6			
3:584			f	7			
3:585			g	3			
3:584			h	8			
3:568			i	9			
3:584	2173:153r		j	9			
3:584	2173:153r		k	9			
3:570–71	2173:128r		l	2			
3:570–71	2173:128r		m	2			
3:568	2173:122v		n	1			
3:568	2173:122v		o	1			
3:569			p	1			
3:575			q	4			
3:650	2173:319–21v		r	5			
3:650	2173:319–21v		s	5			
					9,041	4	0
					1,239	5	6
					10,280	9	6

131

TABLE 5.

| Fiscal Year 1692 | | | Amount Paid | | | | | |
| | | | ARdS Expenses (Categories 4–14) | | | Shared Expenses (Categories 1–3) | | |
Working Year	Purpose of Expenditure	To Whom Paid	livres	sous	deniers	livres	sous	deniers
1691	Expenses of BdR	Clément				1,919	3	5
	Maintenance of PJ	Marchant	105	10	0			
	Rent of 2 buildings occupied by BdR	Abp. of Rouen				5,000	0	0
	Unspecified	Couplet	159	14	6			
	Laboratory	Bourdelin	159	14	0			
1691–92	Maintenance of roof at Obs.	Estienne Yvon, roofer				3,135	0	0
	Masonry at Obs.	Jean Benoit, contractor				253	19	11
	Maintenance of windows at BdR	Jean Gombault				100	0	0
	Maintenance of glass at Obs.	La veuve Janson				500	0	0
	Woodwork at Obs.	Laurens Rochebois				187	19	3
	Small expenses of Obs., '91–June '92	Couplet	220	13	0			
1692	Drawings of plants for AdS	Joubert	300	0	0			
	27 pages of drawings for the new natural history of animals	Chastillon	319	0	0			
	Porterage of dead ostrich & civet from the Versailles menagerie to BdR, then to JR for Du Verney	Camille Le Tellier, for BdR	4	9	0			
	Unspecified, Jan.–June '92	Couplet, concierge of Obs.	143	1	0			
	Drawings & engravings of plants	Joubert, master painter & miniaturist				600	0	0
	36 drawings of rare plants in continuation of miniatures for CdR	Joubert				800	0	0
	Retainer for mathematical instruments	Gosselin & Lagny	200	0	0			
	Retainer for pendulum clocks	Thuret	300	0	0			
	Porter at Obs., salary & livery	Voirie	230	0	0			
	Total Fiscal Year 1692		2,142	1	6	12,496	2	7

[a] Mar. 1692, GC, DA. For the remainder of the 1690s reimbursements for the small or extraordinary expenses of the BdR, which may have included some expenditures on behalf of the Academy, were paid to Clément. A few such entries have been spotted in the records of the PC: AN G⁷ 898 (Apr., Dec. 1695), 899 (July, Oct., Dec. 1696), 901 (May 1697, Jan. 1698). These quarterly payments range in amount from 317 lv. to 961 lv. 6 s.

[b] May 1692, PC, DA.

[c] 2 June 1692, PC, DA; BN MS. shows 160 lv. due.

[d] July 1692, PC, DA.

[e] May 1692, PC, DA.

[f] Nov. 1692, PC, DA.

[g] Dec. 1692, PC, DA.

Continued

Source			Notes	Category	Total		
CdB	AN G[7]	BN MS.			livres	sous	deniers
	894		[a]	1			
3:730				7			
3:709				3			
	894		[b]	8			
	894	n. a. fr. 5147	[c]	6			
3:698–99				1			
3:714				1			
3:719				1			
3:719				1			
3:718				1			
3:723				8			
	894		[d]	4			
	894		[e]	11			
		Arch. A. R. 2: 1r; Clairambault 814: 643–45		10			
	894		[f]	8			
	894		[g]	2			
3:720				2			
3:798				5			
3:798				5			
3:787				9			
					14,638	4	1

TABLE 5.

Fiscal Year 1693		To Whom Paid	Amount Paid					
			ARdS Expenses (Categories 4–14)			Shared Expenses (Categories 1–3, 7*)		
Working Year	Purpose of Expenditure		livres	sous	deniers	livres	sous	deniers
1692	Maintenance of PJ, & of amphitheater & terrace of PJ	Marchant				115	0	0
	Rent of 2 buildings occupied by BdR	Abp. of Rouen				5,000	0	0
	Unspecified	Homberg, Chastillon, & Colson	892	0	0			
1692–93	Maintenance of glass at BdR	Gombault				281	17	10
	Maintenance of windows at Obs.	La veuve Janson				400	0	0
	Work, beyond maintenance, on glass at Obs.	La veuve Janson				68	0	8
	Maintenance of roof at Obs.	Yvon				2,190	0	0
	Small repairs at Obs.	Couplet	74	15	0			
	Laboratory, '92–June '93	Bourdelin	203	8	0			
1693	30 drawings of plants in continuation of miniatures for CdR	Joubert				750	0	0
	Retainer for pendulum clocks	Thuret	300	0	0			
	Retainer for mathematical instruments	Gosselin & Lagny	200	0	0			
	Laboratory	Homberg	523	4	0			
	Porter at Obs., salary & livery	Voirie	230	0	0			
	Laboratory	Unknown	133	10	0			
	Total Fiscal Year 1693		2,556	17	0	8,804	18	6

[a] Jan. 1693, PC, DA.

[b] Dec. 1696–Mar. 1697, GC; no DA after May 1695. BN MS. gives amount due as 203 lv. 8 s. 6 d.

[c] 2 separate items.

[d] *Estat* of 20 July 1693; Aug. 1695, GC.

[e] Not found in AN G⁷; Bourdelin states that this sum was paid by Gruyn on 13 Dec. 1696.

Continued

	Source				Total		
CdB	AN G⁷	BN MS.	Notes	Category	livres	sous	deniers
3:863				7*			
3:867				3			
	895		ª	8			
3:850				1			
3:851				1			
3:851				1	·		
3:842				1			
3:855				9			
	895	n. a. fr. 5147	ᵇ	6			
3:852			ᶜ	2			
3:935				5			
3:934–35				5			
	895		ᵈ	6			
3:920				9			
		n. a. fr. 5147:130r	ᵉ	6			
					11,361	15	6

TABLE 5.

Fiscal Year 1694		To Whom Paid	Amount Paid					
			ARdS Expenses (Categories 4–14)			Shared Expenses (Categories 1–3, 7*)		
Working Year	Purpose of Expenditure		livres	sous	deniers	livres	sous	deniers
1682–83	Woodwork at Obs. & BdR	Claude Tannevot & Jean Mauger, carpenters				5,074	17	4
1693	Rent of 2 buildings occupied by BdR	Abp. of Rouen				5,000	0	0
	Maintenance of and repairs to roof of Obs.	Yvon				1,460	0	0
	Small expenses of and repairs to Obs.	Couplet	73	14	0			
1693–94	Maintenance of PJ, & of amphitheater & terrace of PJ	Marchant				171	10	0
	Maintenance of windows at BdR	Gombault				375	12	7
	Maintenance of windows at Obs., & other work	La veuve Janson				424	4	4
1694	Rent of 2 buildings occupied by BdR	Abp. of Rouen				5,000	0	0
	Porter at Obs., salary	Voirie	200	0	0			
	10 drawings of rare plants in continuation of miniatures for CdR	Joubert				250	0	0
	Retainer to maintain pendulum clocks	Thuret	300	0	0			
	Laboratory	Homberg	513	10	0			
	Total Fiscal Year 1694		1,087	4	0	17,756	4	3

* 2 separate items.
b Pd. Dec. 1694, PC, DA.

Continued

	Source				Total		
CdB	AN G[7]	BN MS.	Notes	Category	livres	sous	deniers
3:1005				1			
3:1001				3			
3:982–83				1			
3:992				9			
3:997				7[*]			
3:988–89				1			
3:988			[a]	1			
3:1008				3			
3:1061				9			
3:990				2			
3:1076				5			
	897		[b]	6			
					18,843	8	3

TABLE 5.

| | Fiscal Year 1695 | | Amount Paid | | | | | |
| | | | ARdS Expenses (Categories 4–14) | | | Shared Expenses (Categories 1–3, 5*, 7*) | | |
Working Year	Purpose of Expenditure	To Whom Paid	livres	sous	deniers	livres	sous	deniers
1678–83	Work on glass at BdR	Heirs of Jacquet				3,419	11	3
1694	Porter at Obs., livery	Voirie	30	0	0			
	Due to porter of Obs., for salary and work on door of Obs.	La veuve Voirie	90	0	0			
	Rent of 2 buildings occupied by BdR	Abp. of Rouen				5,000	0	0
	Laboratory	Bourdelin	147	8	0			
	Laboratory	Unknown	55	10	0			
	Unspecified	Du Verney	159	15	0			
	Woodwork at Obs.	Rosier	6	15	0			
1694?	Unspecified	Couplet	104	0	0			
1694–95	Maintenance of windows & other work at BdR	Gombault				207	9	3
	Maintenance of glass at Obs.	La veuve Janson				400	0	0
1695	Porter at Obs.	Contat	120	0	0			
	Retainer for maintenance of pendulum clocks at all royal buildings in Paris & Versailles	Jacques Thuret, clockmaker				300	0	0
	Laboratory	Homberg	1,122	3	0			
	Maintenance of roof at Obs.	Yvon				3,203	19	2
1695?	Several drawings	Joubert	246	10	0			
	Total Fiscal Year 1695		2,082	1	0	12,530	19	8

ᵃ This completes payment of 5,879 lv. 11 s. 3 d.

ᵇ Work on the door cost 10 lv.

ᶜ The entry also includes *taxations* of 41 lv. 13 s. 4 d.

ᵈ Mar. 1695, PC, DA.

ᵉ No confirmation in AN G⁷. Bourdelin wrote: "Le 18 octobre 1695 Monsieur Simon secretaire de Monsieur labbé Bignon sest donné la peine de maporter les 55 livres 10 sous sans demander de quittance."

ᶠ Mar. 1695, PC, DA.

ᵍ Mar. 1695, PC, DA.

ʰ 2 separate items.

ⁱ Duties but not pay were increased, probably on the accession of Isaac Thuret's son to these responsibilities.

ʲ Quarterly payments, Apr., Aug., Oct., Dec. 1695, PC, DA.

ᵏ Yvon was paid at the rate of 2,920 lv. a year to maintain the roofs of royal buildings, including the Obs., in Paris.

ˡ Oct. 1695, PC, DA.

Continued

Source			Notes	Category	Total		
CdB	AN G⁷	BN MS.			livres	sous	deniers
3:1137			a	1			
3:1122				9			
3:1192			b	9			
3:1138			c	3			
	898	n. a. fr. 5147	d	6			
		n. a. fr. 5147:132v	e	6			
	898		f	10			
3:1118				9			
	898		g	8			
3:1120			h	1			
3:1119				1			
3:1192				9			
3:1202			i	5*			
	898		j	6			
3:1120, 1146, 1148			k	1			
	898		l	4			
					14,613	0	8

139

TABLE 5.

Fiscal Year 1696		To Whom Paid	Amount Paid					
			ARdS Expenses (Categories 4–14)			Shared Expenses (Categories 1–3, 5*, 7*)		
Working Year	Purpose of Expenditure		livres	sous	deniers	livres	sous	deniers
1694–95	Metalwork at Obs.	Nicolas Raguin, metalworker	47	3	0			
1695	Maintenance & cultivation of PJ, & cleaning of amphitheater	Jean Brément, gardener of JR				109	0	0
1695?	Drawings & maps	Chastillon	300	0	0			
	24 drawings of rare plants in continuation of book of miniatures for CdR	Joubert				600	0	0
1695–96	Maintenance of windows and other work at Obs.	La veuve Janson				413	18	4
	Maintenance of windows and other work at BdR	Gombault				274	14	8
1696	6 large telescope lenses for Obs.	Hartsoeker	420	0	0			
	Laboratory	Homberg & other unnamed persons	1,322	6	0			
	Retainer for maintenance of pendulum clocks	Thuret				300	0	0
	Porter at Obs., salary	Contat	180	0	0			
	Porter at Obs., bonus	Contat	100	0	0			
	Total Fiscal Year 1696		2,369	9	0	1,697	13	0

[a] Oct. 1696, PC, DA.
[b] 2 separate items.
[c] 2 separate items.
[d] July 1696, PC, DA.
[e] Quarterly payments, Feb., May, July, Dec. 1696, PC, DA.

Continued

	Source				Total		
CdB	AN G⁷	BN MS.	Notes	Category	livres	sous	deniers
4:45				9			
4:54				7•			
	899		•	12			
4:47				2			
4:46			b	1			
4:46–47			c	1			
	899		d	5			
	899		•	6			
4:135				5•			
4:125				9			
4:273				9			
					4,067	2	0

TABLE 5.

			Amount Paid					
			ARdS Expenses (Categories 4–14)			Shared Expenses (Categories 1–3, 5*, 7*)		
Fiscal Year 1697								
Working Year	Purpose of Expenditure	To Whom Paid	livres	sous	deniers	livres	sous	deniers
1694–97	Experiments on minerals	Morin	600	0	0			
1695–96	Maintenance of the grande terrasse & other parts of Obs.	Couplet	69	2	0			
1696	Laboratory	Bourdelin	168	7	0			
	Laboratory	Homberg	346	12	0			
	Copyist	Du Hamel	150	0	0			
	Maintenance & cultivation of PJ of JR, and cleaning the amphitheater	Brément				100	0	0
	Porter at Obs., 2 bonuses	Contat	250	0	0			
	Rent of 2 buildings occupied by BdR	Abp. of Rouen				5,000	0	0
1696?	Unspecified	Couplet	117	0	0			
	Unspecified	Du Verney	145	11	0			
1696–97	24 drawings of rare plants for CdR	Joubert				600	0	0
	Maintenance of windows and other work at Obs.	La veuve Janson				484	7	8
	Maintenance of windows & other work at BdR	Gombault				382	1	8
	Porter at Obs., salary & livery	Contat	230	0	0			
1697	Unspecified	Couplet	432	9	0			
	Unspecified	Du Verney	250	18	6			
	Small expenses	Tournefort	118	2	0			
	Laboratory	Bourdelin	165	0	0			
	Laboratory	Homberg	1,190	2	0			
	Retainer to maintain pendulum clocks	Thuret				300	0	0
	Total Fiscal Year 1697		4,233	3	6	6,866	9	4

[a] Jan. 1698, PC; no DA for 1698. The entry reads: "Au Sieur Morin pour son remboursement de la depense qu'il a faite en experience sur les mineraux pendant 1694, 1695, 1696, et 1697, ordonnance du dernier decembre 1697. . . ."

[b] Mar. 1697, PC, DA. Bourdelin reported that Homberg kindly brought him the money.

[c] Mar. 1697, PC, DA, for the last quarter of 1696.

[d] Mar. 1697, PC, DA, along with monies to Bourdelin and Homberg (above), in one lump sum amounting to 664 lv. 19 s.

[e] Approved in fiscal year 1696, paid to the CdB in Feb. 1697 (GC).

[f] May 1697, PC, DA, in lump sum to Du Verney, Couplet, and others totaling 682 lv. 1 s.

[g] May 1697, PC, DA, in lump sum to Du Verney, Couplet, and others totaling 682 lv. 1 s.

[h] 2 separate items.

[i] 2 separate items.

[j] 2 separate payments.

[k] Jan. 1698, PC.

[l] Jan. 1698, PC.

[m] Jan. 1698, PC.

[n] Jan. 1698, PC; Bourdelin asked Simon, Bignon's secretary, for reimbursement at the beginning of Jan. 1698 and received it that same month. But he requested 165 lv. 10 s.

[o] Quarterly payments, May, Aug., Oct. 1697, PC, DA, Feb. 1698, PC.

Continued

Source			Notes	Category	Total		
CdB	AN G⁷	BN MS.			livres	sous	deniers
	901		a	13			
4:191–92				9			
	901	n. a. fr. 5147:134r	b	6			
	901		c	6			
	901		d	8			
4:198				7*			
4:273, 277				9			
4:201–2	899		e	3			
	901		f	8			
	901		g	10			
4:187			h	2			
4:186			i	1			
4:186–87			j	1			
4:272				9			
	901		k	8			
	901		l	10			
	901		m	14			
	901	n. a. fr. 5147:135r	n	6			
	901		o	6			
4:271				5*			
					11,099	12	10

143

TABLE 5.

			Amount Paid					
Fiscal Year 1698			ARdS Expenses (Categories 4–14)			Shared Expenses (Categories 1–3, 5*, 7*)		
Working Year	Purpose of Expenditure	To Whom Paid	livres	sous	deniers	livres	sous	deniers
1680	Woodwork at Obs.	Jacques, carpenter	59	6	8			
1695–97	Metalwork at Obs.	Raguin, locksmith	50	5	6			
1697	Small repairs at Obs.	Couplet	24	17	0			
	Maintenance and cultivation of PJ of JR, and for other work outside his normal responsibilities	Brément				100	0	0
	Rent of 2 houses for BdR	Abp. of Rouen				5,000	0	0
1697–98	24 drawings of rare plants for CdR	Joubert				600	0	0
	Maintenance of windows & other work at Obs.	La veuve Janson & Antoine-Charles Janson				437	11	3
	Maintenance of windows & other work at BdR	Gombault				450	12	1
	Porter at Obs., salary, livery, & bonus	Contat	380	0	0			
1697–98?	Scientific instruments & machines	Couplet	370	7	0			
	Machines & tools	Obry	879	17	0			
	Small expenses for the experiments of AdS	Couplet	158	8	0			
	Various expenses of AdS	Fontenelle	609	5	0			
	Laboratory	Bourdelin	191	2	0			
1698	Retainer to maintain pendulum clocks	Thuret				300	0	0
	Scale model of the Samaritaine machine & pump which he made	Couplet	400	0	0			
	Laboratory	Homberg	1,172	12	0			
	Total Fiscal Year 1698		4,296	0	2	6,888	3	4
			−191	2	0			
	Adjusted Total		4,104	18	2			

ᵃ 2 separate payments.

ᵇ 2 separate items.

ᶜ 2 separate items.

ᵈ 2 separate items; working year for one uncertain.

ᵉ Aug. 1698, PC; no DA exists for this year. The entry reads: "Au Sieur Couplet tant pour reparations faites aux machines de l'observatoire, que pour une pendule fournie au Sieur Couplet son fils, pour les observations quil a eu ordre de faire en Portugal."

ᶠ June 1698, PC.

ᵍ June 1698, PC.

ʰ Jan. 1699, PC.

ⁱ Bourdelin said he received this sum from Fontenelle on 28 Jan. 1699. There is no entry in the accounts of the treasury either in Bourdelin's name or in this amount. This sum is probably included in the 609 lv. 5 s. due to Fontenelle, 27 Jan. 1699, PC. Hence the adjusted total for this year.

ʲ Quarterly payments, June, Aug., Dec. 1698, Jan. 1699, PC.

Continued

	Source				Total		
CdB	AN G[7]	BN MS.	Notes	Category	livres	sous	deniers
4:333				9			
4:334				9			
4:340				9			
4:344				7[*]			
4:348				3			
4:335, 336			[a]	2			
4:335			[b]	1			
4:335			[c]	1			
4:416, 421			[d]	9			
	902		[e]	5			
	902		[f]	5			
	902		[g]	8			
	902		[h]	8			
		n. a. fr. 5147:136r	[i]	6			
4:430				5[*]			
4:431				5			
	902		[j]	6			
					11,184	3	6
					−191	2	0
					10,993	1	6

TABLE 5.

Fiscal Year 1699			Amount Paid					
			ARdS Expenses (Categories 4–14)			Shared Expenses (Categories 1–3, 5*, 7*)		
Working Year	Purpose of Expenditure	To Whom Paid	livres	sous	deniers	livres	sous	deniers
1697–98	Repairs to Obs.	Couplet	204	15	0			
1698	Maintenance of glass at Obs.	La veuve Janson & A.-C. Janson				400	0	0
	Rent of buildings occupied by BdR	Abp. of Rouen				5,000	0	0
1698–99	Maintenance of glass at BdR	Gombault				200	0	0
	24 drawings of rare plants for CdR	Joubert				750	0	0
	Payment on account for 400 drawings of rare plants	Joubert				6,600	0	0
	For polishing the burning mirror made by La Garouste	Menard	110	0	0			
	Porter of Obs., salary & livery, 1 Oct. '98–26 Mar. '99	Contat	127	0	0			
1699	Porter of Obs., salary, 27 Mar.–30 June	Barabel	52	15	0			
	Retainer to maintain pendulum clocks	Thuret				300	0	0
	Unspecified	Couplet	7,530	2	0			
	Unspecified	Couplet	5,119	18	0			
	Unspecified	Homberg	608	10	0			
	Unspecified	Marchant	363	10	0			
	Unspecified	Fontenelle	500	0	0			
	Unspecified	Simonneau	3,097	0	0			
	Total Fiscal Year 1699		17,713	10	0	13,250	0	0

[a] AN G⁷ 902 confirms payment into the buildings account, Dec. 1698, GC.

[b] Listed in separate *ordonnances à délivrer à M. des Granges,* to whom the *ordonnances* to pay academicians' pensions were also made out. These *ordonnances* were recorded in the following weeks: 3, 14 Mar., 9, 30 May; 21 July; 12 Sept.; 5, 29 Dec. 1699; 26 Jan. 1700. They were in the following amounts: 104 lv. 1 s.; 450 lv. 6 s.; 652 lv. 10 s.; 948 lv.; 478 lv. 4 s.; 743 lv. 4 s.; 965 lv. 10 s.; 960 lv.; 993 lv. 17 s.; 308 lv.; 926 lv. 10 s.; 652 lv. 10 s.; 104 lv. 1 s. AN G⁷ 903 confirms payments from January 1700.

[c] Listed in separate *ordonnances à délivrer à M. Couplet* or to "Couplet luy-même." These were recorded on 28 Mar. and 17 Oct. 1699 and were in the amounts of 3,600 and 573 lv., and 946 lv. 18 s. It is not certain that this Couplet is the academician.

[d] 12 Sept., 5 Dec. 1699, *ordonnances à délivrer à M. des Granges.* AN G⁷ 903 confirms payments from January 1700.

[e] 12 Sept. 1699, 26 Jan. 1700, *ordonnances à délivrer à M. des Granges.* AN G⁷ 903 confirms payments from January 1700.

[f] 26 Jan. 1700, *ordonnance à délivrer à M. des Granges.* AN G⁷ 903 confirms payments from January 1700.

[g] 21 July, 5, 29 Dec. 1699, *ordonnances à délivrer à M. des Granges.* The *ordonnance* of 29 Dec. for 900 lv. may be confirmed in AN G⁷ 903.

Continued

Source					Total		
CdB	AN G⁷	BN MS.	Notes	Category	livres	sous	deniers
4:482				9			
4:476				1			
4:492	902		ᵃ	3			
4:476–77				1			
4:477				2			
4:477				2			
4:484				5			
4:557, 561				9			
4:558				9			
4:569				5ᵃ			
	903, 904		ᵇ	8			
	903, 904		ᶜ	8			
	903, 904		ᵈ	6			
	903, 904		ᵉ	14			
	903, 904		ᶠ	8			
	903, 904		ᵍ	12			
					30,963	10	0

TABLE 6. Summary of Research Expenses of the Academy

Fiscal Year	Ministerial Protector	Shared Expenses																AdS Expenses						
		1 Physical Plant & Small Expenses			2 Illustrations of Plants for CdR			3 Rent			5* Scientific Instruments			7* Petit Jardin			Total Shared Expenses			4 Engravings & Drawings of Plants			5 Scientific Instruments & Models		
		livres	sous	deniers	livres	sous	deniers	livres	sous	deniers	livres	sous	deniers	livres	sous	deniers	livres	sous	deniers	livres	sous	deniers	livres	sous	deniers
1690	Louvois	4078	18	11	600	0	0	5000	0	0							9678	18	11	726	0	0	600	0	0
1691	Louvois	2152	14	0	600	0	0	5000	0	0							7752	14	0	528	0	0			
1691	Pontchartrain	296	15	6	300	0	0										596	15	6				500	0	0
1692	Pontchartrain	6096	2	7	1400	0	0	5000	0	0							12496	2	7	300	0	0	500	0	0
1693	Pontchartrain	2939	18	6	750	0	0	5000	0	0				115	0	0	8804	18	6				500	0	0
1694	Pontchartrain	7334	14	3	250	0	0	10000	0	0				171	10	0	17756	4	3				300	0	0
1695	Pontchartrain	7230	19	8				5000	0	0	300	0	0				12530	19	8	246	10	0			
1696	Pontchartrain	688	13	0	600	0	0				300	0	0	109	0	0	1697	13	0				420	0	0
1697	Pontchartrain	866	9	4	600	0	0	5000	0	0	300	0	0	100	0	0	6866	9	4						
1698	Pontchartrain	888	3	4	600	0	0	5000	0	0	300	0	0	100	0	0	6888	3	4				1650	4	0
1699	Pontchartrain	600	0	0	7350	0	0	5000	0	0	300	0	0				13250	0	0				110	0	0
	Total	33173	9	1	13050	0	0	50000	0	0	1500	0	0	595	10	0	98318	19	1	1800	10	0	4580	4	0

Note: On the assumption that 191 lv. 2 s. of the 690 lv. 5 s. paid to Fontenelle was earmarked for Bourdelin, I have reduced the total of small expenses (category 8) for this year by 191 lv. 2 s.

	AdS Expenses (Continued)								
6 Chemical Laboratory	7 Petit Jardin	8 Small Expenses	9 Observatory	10 Anatomical Research	11 Engravings & Drawings of Animals	12 Other Engravings & Drawings	13 Research on Minerals	14 Research on Plants	Total AdS Expenses
livres sous deniers	livres sous deniers	livres sous deniers	livres sous deniers	livres sous deniers	livres sous deniers	livres sous deniers	livres sous deniers	livres sous deniers	livres sous deniers
1019. 4. 6	89. 0. 0	65. 17. 6	480. 0. 0	316. 10. 0					3296. 12. 0
240. 7. 0	63. 4. 0	180. 9. 0	276. 10. 0						1288. 10. 0
			142. 10. 0						642. 10. 0
159. 14. 0	105. 10. 0	523. 8. 6	230. 0. 0	4. 9. 0	319. 0. 0				2142. 1. 6
860. 2. 0		892. 0. 0	304. 15. 0						2556. 17. 0
513. 10. 0			273. 14. 0						1087. 4. 0
1325. 1. 0		104. 0. 0	246. 15. 0	159. 15. 0					2082. 1. 0
1322. 6. 0			327. 3. 0			300. 0. 0			2369. 9. 0
1870. 1. 0		699. 9. 0	549. 2. 0	396. 9. 6			600. 0. 0	118. 2. 0	4233. 3. 6
1363. 14. 0		576. 11. 0ª	514. 9. 2						4104. 18. 2
608. 10. 0		13150. 0. 0	384. 10. 0			3097. 0. 0		363. 10. 0	17713. 10. 0
9282. 9. 6	257. 14. 0	16191. 15. 0	3729. 8. 2	877. 3. 6	319. 0. 0	3397. 0. 0	600. 0. 0	481. 12. 0	41516. 16. 2

TABLE 7. Allocation of Research

Composite Categories	Individual Categories		Totals		% of Total	
			Individual Categories	Composite Categories	Individual	Composite
	No.	Name	lv. s. d.	lv. s. d.	Categories	Categories
Natural Philosophy	4	Engravings & Drawings of Plants	546. 10. 0		3%	
	6	Chemical Laboratory	7,414. 8. 0		38.6%	
	7	*Petit Jardin*	105. 10. 0		.5%	
	10	Anatomical Research	560. 13. 6		3%	
	11	Engravings & Drawings of Animals	319. 0. 0		1.7%	
	13	Research on Minerals	600. 0. 0		3.1%	
	14	Research on Plants	118. 2. 0		.6%	
	Total Natural Philosophy			9,664. 3. 6		50.5%
Mathematical Sciences	5	Scientific Instruments & Models	3,870. 4. 0		20%	
	9	Observatory	2,588. 8. 2		13%	
	12	Other Engravings & Drawings	300. 0. 0		2%	
	Total Mathematical Sciences			6,758. 12. 2		35%
Small Expenses	8	Small Expenses	2,795. 8. 6		14.5%	
	Total Small Expenses			2,795. 8. 6		14.5%
Totals		Total Individual Categories	19,218. 4. 2		100%	
		Total Composite Categories		19,218. 4. 2		100%

Expenditure under Pontchartrain

1699				1691–1699			
Totals		% of Total		Totals		% of Total	
Individual Categories lv. s. d.	Composite Categories lv. s. d.	Individual Categories	Composite Categories	Individual Categories lv. s. d.	Composite Categories lv. s. d.	Individual Categories	Composite Categories
—		—		546. 10. 0		1.5%	
608. 10. 0		3.4%		8,022. 18. 0		21.7%	
—		—		105. 10. 0		.3%	
—		—		560. 13. 6		1.5%	
—		—		319. 0. 0		1%	
—		—		600. 0. 0		2%	
363. 10. 0		2%		481. 12. 0		1%	
	972. 0. 0		5.4%		10,636. 3. 6		29%
110. 0. 0		1%		3,980. 4. 0		11%	
384. 10. 0		2%		2,972. 18. 2		8%	
3,097. 0. 0		17.4%		3,397. 0. 0		9%	
	3,591. 10. 0		20.4%		10,350. 2. 2		28%
13,150. 0. 0		74.2%		15,945. 8. 6		43%	
	13,150. 0. 0		74.2%		15,945. 8. 6		43%
17,713. 10. 0		100%		36,931. 14. 2		100%	
	17,713. 10. 0		100%		36,931. 14. 2		100%

TABLE 8. Analysis of Direct Expenditure on the Academy of Sciences, 1666–99

Category	Ministerial Protectors of Academy	Total Spent on the Category			The Category as % of Total Spent On:			Average Spent Yearly
					This Category 1666–99	All Direct Expenditure 1666–99	All Direct Expenditure by the Minister in Question	
A. Pensions to Academicians & Their Assistants[a]		lv.	s.	d.	%	%	%	lv.
	Colbert, 1666–83 (18 yrs.)	645,308.	6.	8	59	30.2	41	35,850
	Louvois, 1684–91 (8 yrs.)	168,833.	6.	8	15	7.9	72	21,104
	Pontchartrain, 1691–99 (9 yrs.)	286,016.	13.	4	26	13.3	88.5	31,780
	Total, 1666–99 (34 yrs.)	1,100,158.	6.	8	100%	51.4		32,358
B. Observatory	Colbert	713,704.	3.	11	98	33.4	45	39,650
	Louvois	12,335.	7.	1	1.6	.6	5	1,542
	Pontchartrain	2,972.	18.	2	.4	.1	1	330
	Total	729,012.	9.	2	100%	34.1		21,442
C. Research	Colbert	223,113.	16.	10	72	10.4	14	12.395
	Louvois	53,025.	5.	9	17	2.5	23	6,628
	Pontchartrain	33,958.	16.	0	11	1.6	10.5	3,773
	Total	310,097.	18.	7	100%	14.5		9,121
						100%		

		Total Spent on All Categories			Each Minister's Expenditure as % of Total Direct Expenditure by All Ministers, 1666–99	Average Spent Yearly on All Categories
D. Summary: Total Direct Expenditure		lv.	s.	d.	%	lv.
	Colbert	1,582,126.	7.	5	74	87,896
	Louvois	234,193.	19.	6	11	29,274
	Pontchartrain	322,948.	7.	6	15	35,883
	Total	2,139,268.	14.	5	100%	62,920

[a] Colbert spent 614,525 lv. on pensions for academicians, and 30,783 lv. 6 s. 8 d. for assistants to the Academy. Louvois spent 159,500 lv. on pensions for academicians, and 9,333 lv. 6 s. 8 d. for assistants to the Academy. These figures are based on *CdB*. Pontchartrain spent 269,550 lv. on pensions for academicians, and 16,446 lv. 13 s. 4 d. for assistants to the Academy (table 2).

BIBLIOGRAPHY

PRIMARY SOURCES: *Manuscripts*

Paris. Archives de l'Académie des Sciences. (AdS).
    Cartons 1666–1793, nos. 1–3: Papers and notebooks from 1666–99, including Bourdelin's laboratory notebooks and Dodart's analyses, called *abrégés*, of Bourdelin's experiments.
    Carton 1680–99: Papers of various academicians, including Sédileau, Morin, Dalesme, and Des Billettes.
    Dossiers: "Pierre Couplet," "Gilles Filleau Des Billettes," "Denis Dodart," "Jacques Jaugeon," "Morin de Toulon," "Père Sébastien Truchet."
    Registre des procès-verbaux des séances. Vols. 1–18 (1666–99). (AdS, Reg.).
Paris. Archives Nationales. (AN).
    G^7 891–904: Trésor royal, feuilles mensuelles et états annuels, 1690–99. See appendix A.
    G^7 973: Trésor royal, appointements, pensions, aumônes, 1694–1712. See appendix A.
    G^7 980–87: Trésor royal, états de distribution, paiemens ordonnez, assignations ordonnez, 1689–1701. See appendix A.
    G^7 991–1000: Trésor royal, pièces justificatives des états de distribution, 1689–1703. See appendices A and B.
    KK 355: Abregé des registres du roy pendant le ministère de Monsieur Colbert, contrôleur général des finances. Parts of the manuscript have been published in *Correspondance des contrôleurs généraux*, 1.
    M 802: Papers pertaining to the history of the Imprimerie royale.
    M 849, 851: Papers of père Sébastien Truchet and Gilles Filleau Des Billettes.
    O^1 656, no. 1: "Gratifications aux gens de lettres, 1687."
    O^1 1934^B 14: Records of accounts payable, mostly eighteenth century, but including an *estat* for the Academies of Sciences and Inscriptions.
    O^1 2173–74: Comptes des bâtiments, 1691 (published in *CdB*).
    PP 151: Naturalités, octrois, pensions, 1635–1742.
Paris. Bibliothèque Nationale. (BN).
    Archives de l'ancien régime, 1 and 2: Expenses of the Bibliothèque du roi, 1680s–90s.
    Clairambault 566: Bignon's papers about the Academies. Fols. 251–52 have been published in Saunders, *Decline and Reform*, 256–59.
    Clairambault 814, fol. 633r-v: History of the Academy of Sciences.
    Fr. 7750: Papers on the finances of France during the seventeenth and eighteenth centuries, including "Abregé historique sur les finances jusqu'au temps de la regence 1715," by abbé Pegere, fols. 1–41; "Memoire presenté au roy par M. le Pelletier après avoir quitté la finance pour lui rendre compte de son administration," fols. 72–87 (published in *Correspondance des contrôleurs généraux*, 1: 554–57).
    Fr. 7801: Papers of the surintendants des bâtiments du roi, 1667–1739.
    Fr. 13070: Papers seized from père Léonard relating to the Academies of Medicine and Sciences.
    Fr. 22225: Bignon's papers about the Academies.
    N. a. fr. 5133–49: Notebooks from chemical laboratory of the Academy of Sciences, kept by Claude Bourdelin, 1666–99.

PRIMARY SOURCES: *Printed*

Académie royale des sciences. *Histoire de l'Académie royale des sciences, avec les mémoires de mathématique et de physique. Tirés des registres de cette Académie (1699–1790)*. Paris: Imprimerie royale, 1702–97. [*Histoire . . . (date)* or *Mémoires . . . (date)*].
——. *Histoire et mémoires de l'Académie royale des sciences depuis 1666 jusqu'à 1699*. 11 vols. Paris: Martin, Coignard, Guerin & La Compagnie des Libraires, 1729–33. (*Histoire* or *Mémoires*).

153

Bernoulli, Johann. *Der Briefwechsel*, ed. O. Spiess. Basel: Birkhäuser Verlag, 1955.

Blegny, Nicolas de. *Le livre commode des adresses de Paris pour 1692 par Abraham du Pradel*, ed. Édouard Fournier. 2 vols. Paris: Paul Daffis, 1878.

Boislisle, A. M. de, ed. See *Correspondance des contrôleurs généraux.*

———, ed. See Saint-Simon, *Mémoires.*

Boze, Claude Gros de. See Gros de Boze, Claude.

Brice, Germain. *Description nouvelle de ce qu'il y a de plus remarquable dans la ville de Paris.* 2 vols. Paris: Nicolas Le Gras, 1684.

Cassini, Jean-Dominique, I. *Anecdotes de la vie . . . rapportées par lui-même*, in J.-D. Cassini, comte de Cassini, *Mémoires pour servir à l'histoire des sciences et à celle de l'Observatoire royale de Paris.* Paris: Bleuet, 1810.

———. *Le neptune françois.* See Pène, Charles de.

———. *Observations astronomiques faites en Hollande, et en Angleterre, en 1697. & 1698.,* in *Mémoires,* 7, pt. 2: 535–72.

——— and Cassini, Jean-Dominique, II. *Observations astronomiques faites en France et en Italie en 1694. 1695. & 1696.,* in *Mémoires,* 7, pt. 2: 461–533.

[Chamillart, Michel de]. "État auquel M. de Chamillart a trouvé les finances du roi, le 6 septembre 1699," ed. abbé Esnault, in "Documents relatifs à l'histoire des finances sous le règne de Louis XIV." See Esnault.

Chapelain, Jean. *Lettres,* ed. Ph. Tamizey de Larroque. 2 vols. Paris: Imprimerie nationale, 1880–83.

Colbert, Jean-Baptiste. *Lettres, instructions et mémoires,* ed. Pierre Clément. 8 vols. Paris: Imprimerie impériale, Imprimerie nationale, 1861–70, 1882.

Colletet, François. *Le journal de Colletet: premier petit journal parisien (1676),* ed. Arthur Heulhard. Paris: Le Moniteur du Bibliophile, 1878.

Les comptes des bâtiments du roi sous le règne de Louis XIV, ed. Jules Guiffrey. 5 vols. Paris: Imprimerie nationale, 1881–1901. (CdB).

La connoissance des temps, ed. Jean Le Febvre. Paris: Estienne Michallet, 1692, 1694–99.

Correspondance des contrôleurs généraux des finances avec les intendants des finances, ed. A. M. de Boislisle. 3 vols. Paris: Imprimerie nationale, 1874–97.

Correspondance administrative sous le règne de Louis XIV entre le Cabinet du Roi, les Secrétaires d'État, le Chancelier de France . . . , ed. G. B. Depping. 4 vols. Paris: Imprimerie nationale, 1850–55.

Couplet, Pierre. "Extrait de quelques lettres écrites de Portugal et du Brésil . . . à Monsieur l'abbé Bignon," in *Mémoires . . . 1700,* 172–78.

Depping, G. B., ed. See *Correspondance administrative sous le règne de Louis XIV.*

Des Billettes, Gilles Filleau. "Description d'une nouvelle manière de porte d'écluse qu'on a pratiquée dans l'entreprise de la nouvelle navigation de la Seyne," in *Mémoires . . . 1699,* 63–68.

Du Hamel, Jean-Baptiste. *Regiae Scientiarum Academiae Historia.* Paris: J.-B. Delespine, 1701 [1698]. (Historia).

État de la France, ed. Nicolas Besongne, Louis Trabouillet. Paris: various printers, 1663, 1692, 1694, 1697, 1698, 1699.

Fontenelle, Bernard de. *Éloges des academiciens de l'Académie royale des sciences morts depuis l'an 1699.* Vol. 3 of *Oeuvres diverses de M. de Fontenelle.* New ed. The Hague: Gosse et Neaulme, 1729.

———. *Histoire de l'Académie royale des sciences depuis 1666 jusqu'à 1699.* 2 vols. Paris: Martin, Coignard, Guerin, 1733. Vols. 1–2 of *Académie royale des sciences, Histoire et mémoires . . . depuis 1666 jusqu'à 1699.* (Histoire).

Furetière, Antoine. *Dictionnaire universel.* New ed. 2 vols. The Hague and Rotterdam: Chez Arnout et Reinier Leers, 1694.

———. *Recueil des factums d'Antoine Furetière de l'Académie françoise contre quelques-uns de cette Académie suivi des preuves et pièces historiques données dans l'édition de 1694,* ed. Charles Asselineau. 2 vols. Paris: Poulet-Malassis et de Broise, 1859.

Gallois, abbé Jean. "Extrait du livre intitulé *Observations physiques & mathématiques envoyées des Indes & de la Chine . . . ,*" in *Mémoires,* 10: 130–38.

Gouye, Thomas. *Observations physiques et mathématiques, pour servir à la perfection de l'astronomie et de la géographie. Envoyées des Indes et de la Chine à l'Académie,* in *Mémoires,* 7, pt. 2: 741–875.

———. *Observations physiques et mathématiques, pour servir à la perfection de l'astronomie et de la géographie. Envoyées de Siam à l'Académie,* in *Mémoires,* 7, pt. 2: 605–740.

Gros de Boze, Claude. *Histoire de l'Académie royale des inscriptions et belles lettres depuis son établissement jusqu'à présent avec les Mémoires.* . . . 2 vols. Paris: Imprimerie royale, 1717.

Guiffrey, Jules, ed. See *Les comptes des bâtiments.*

Huygens, Christiaan. *Oeuvres complètes.* 22 vols. Amsterdam: Swets and Zeitlinger, N. V.; The Hague: Martinus Nijhoff, 1888–1950.

Journal des sçavans. Paris and Amsterdam, 1665–99.

La Hire, Philippe de. *Traité des épicycloïdes* (1696), in *Mémoires,* 9: 341–447.

——. *Traité de méchanique* (1694), in *Mémoires,* 9: 1–333.

Le Febvre, Jean. See *La connoissance des temps.*

Leibniz, Gottfried Wilhelm. *Lettres et opuscules inédits* . . . , ed. A. Foucher de Careil. Paris: Librairie Philosophique de Ladrange, 1854.

——. *Oeuvres,* ed. A. Foucher de Careil. 2 vols. Paris: Firmin Didot, 1859–60. (Lettres).

——. *Philosophical Papers and Letters,* trans. and ed. by Leroy E. Loemker. 2 vols. Chicago: The University of Chicago Press, 1956.

Lister, Martin. *A Journey to Paris in the Year 1698* (London, 1699). References are to the 3d ed., ed. R. P. Stearns. Facsimile reprint in *The History of Sciences,* 4. Urbana: University of Illinois Press, 1967.

Louis XIV. *Edit du roy, portant création de douze cens mille livres de rente au denier dix huit sur l'Hostel de la Ville de Paris; et faculté au proprietaires qui ont des rentes au denier vingt de les convertir au denier dixhuit. Donné à Versailles au mois d'avril 1692. Registré en Parlement.* Paris: Estienne Michallet, 1692.

——. *Edit du roy, portant création de rentes viagères à fonds perdu, constituées sur l'Hostel de Ville de Paris, & distribuées en differentes classes sur un pied proportionné à l'âge des rentiers. Donné à Versailles au mois d'aoust 1693. Registré en Parlement le 2 septembre 1693.* Paris: Estienne Michallet, 1693.

Malebranche, Nicolas. *Oeuvres complètes,* ed. André Robinet. 20 vols. Paris: J. Vrin, 1958–78.

Le neptune françois. See Pène, Charles de.

Oldenburg, Henry. *The Correspondence,* ed. A. Rupert Hall and Marie Boas Hall. 10 vols. Madison, Wisc.: University of Wisconsin Press, 1965–75.

Pène, Charles de, Joseph Sauveur, J.-D. Cassini, et al. *Le neptune françois, ou recueil des cartes marines, levées et gravées par ordre du roy. Premier volume. Contenant les costes de l'Europe sur l'Ocean, depuis Dronthem et Norvege jusques au Detroit de Gibraltar, avec la Mer Baltique.* Paris: Imprimerie royale, 1693; Paris: Hubert Jaillot, 1693. Two different printings.

Perrault, Charles. *Mémoires de ma vie* in *Mémoires de ma vie par Charles Perrault. Voyage à Bordeaux (1669) par Claude Perrault,* ed. Paul Bonnefon. Paris: Librairie Renouard, H. Laurens, 1909.

Racine, Jean. *Oeuvres,* ed. Paul Mesnard. 8 vols. Paris: Librairie Hachette, 1887–1921.

Saint-Simon, Louis Rouvroy, duc de. *Mémoires.* New ed., ed. A. de Boislisle. 41 vols. Paris: Hachette, 1879–1928.

Sauveur, Joseph. *Explication des echelles pour les calculs de marine, pour servir d'introduction aux cartes marines gravées par ordre du roy.* Paris: Louis Sevestre, 1692. The two copies of this edition at the BN lack the illustrations of the scale; the illustrations may be seen in the *Neptune françois,* which includes in its introductory materials Sauveur's entire treatise.

——. *Le neptune françois.* See Pène, Charles de.

Simonneau, Louis, père et fils. *Recueil d'estampes pour servir à l'histoire de l'imprimerie, de la gravure, et des arts et métiers, 1694–1719.* This collection contains some of the plates corresponding to the work of the Compagnie des arts et métiers during the 1690s and of the Academy of Sciences in the early eighteenth century. BN, Département des Estampes, Réserve.

Sourches, Louis-François de Bouschet, marquis de. *Mémoires* . . . *sur le règne de Louis XIV,* ed. Gabriel-Jules, le comte de Cosnac, Arthur Bertrand, and Édouard Pontal. 14 vols. Paris: Hachette, 1882–93, 1912.

Tournefort, Joseph Pitton de. *Élemens de botanique, ou methode pour connoître les plantes.* 3 vols. Paris: Imprimerie royale, 1694.

——. *Histoire des plantes qui naissent aux environs de Paris, avec leurs usages dans la médecine.* Paris: Imprimerie royale, 1698.

SECONDARY SOURCES

Académie des Sciences. See Institut de France.

André, Louis and Émile Bourgeois. *Recueil des instructions données aux ambassadeurs et ministères de France depuis les traités de Westphalie jusqu'à la révolution française.* 30 vols. Paris: various publishers, 1884–1983. Vols. 16, 17, 18 constitute the *Recueil . . . Hollande,* ed. André and Bourgeois, 3 vols. (Paris: E. de Boccard, 1922–24). Page references are exclusively to the volumes on Holland, numbered 1, 2, and 3.

Antoine, Michel. *Le conseil du roi sous le règne de Louis XV.* Mémoires et documents publiés par la Société de l'École des Chartes, 19. Geneva: Librairie Droz, 1970.

Avenel, Georges le vicomte d'. *Histoire économique de la propriété, des salaires, des denrées et de tous les prix en général depuis l'an 1200 jusqu'en l'an 1800.* 6 vols. Paris: Ernest Leroux, 1884–1912.

Bailly, M. A. *Histoire financière de la France, depuis l'origine de la monarchie jusqu'à la fin de 1786.* Paris: Moutardier, 1830.

Balteau, J., M. Barroux, M. Prevost, et al. *Dictionnaire de biographie française.* 16 vols. Paris: Librairie Letouzey et Ané, 1933–81. *(DBF).*

Barber, Elinor G. *The Bourgeoisie in 18th Century France.* Princeton: Princeton University Press, 1955.

Bénézit, Émmanuel. *Dictionnaire critique et documentaire des peintres, sculpteurs, dessinateurs et graveurs.* New ed. 10 vols. Paris: Librairie Gründ, 1976.

Berger, Patrice. "French Administration in the Famine of 1693." *European Studies Review* 8 (1978): 101–27.

——. "Pontchartrain and the Grain Trade during the Famine of 1693." *The Journal of Modern History* 48 (December 1976, on-demand supplement): 37–86.

——. "Rural Charity in Late Seventeenth Century France: The Pontchartrain Case." *French Historical Studies* 10 (1978): 393–415.

Bernard, Auguste. *Histoire de l'Imprimerie royale au Louvre.* Paris: Imprimerie impériale, 1867.

Berton, André. *L'impôt de la capitation sous l'ancien régime.* Thèse pour le doctorat, 1907, Université de Paris, Faculté de droit. Paris: Librairie de la Société de Recueil J.-B. Sirey, et du Journal de Paris, 1907.

Bertrand, Joseph L. F. *L'Académie des sciences et les académiciens de 1666 à 1793.* Paris: J. Hetzel, 1869; reprinted, Amsterdam, 1969.

——. "Les Académies d'autrefois. L'ancienne Académie des sciences, par Alfred Maury . . . , 1865.—Procès-verbaux inédits des séances de l'Académie des Sciences." *Journal des savants* (1866): 337–53, 420–32, 576–93, 715–25, 758–69; (1867): 167–82, 752–66; (1868): 107–23.

Bluche, François and Jean-François Solnon. *La véritable hiérarchie sociale de l'ancienne France: Le tarif de la première capitation (1695).* Geneva: Librairie Droz, 1983.

Boislisle, A. M. de. "Les conseils sous Louis XIV," in Saint-Simon, *Mémoires,* 4: 377–439, 5: 437–82, 6: 477–514, 7: 405–43.

Bourgeois, Émile and Louis André. *Les sources de l'histoire de France. XVIIe siècle (1610–1715).* 8 vols. Paris: Auguste Picard, 1913–35.

Braudel, Fernand and Ernest Labrousse, eds. *Histoire économique et sociale de la France.* 4 vols. Paris: Presses Universitaires de France, 1970–82. Vol. 2, *Des derniers temps de l'âge seigneurial aux préludes industriels (1660–1789),* by Ernest Labrousse, Pierre Léon, Pierre Goubert, Jean Bouvier, Charles Carrière, Paul Harsin.

Brown, Harcourt. *Science and the Human Comedy: Natural Philosophy in French Literature from Rabelais to Maupertuis.* Toronto: University of Toronto Press, 1976.

——. *Scientific Organizations in Seventeenth Century France (1620–1680).* History of Science Society Publications, new series 5. Baltimore: Williams and Wilkins Co., 1934.

Brunot, Ferdinand. *Histoire de la langue française des origines à 1900.* 13 vols. in 21. Paris: Librairie Armand Colin, 1905–72.

Brygoo, Édouard. "Les médecins de Montpellier et le Jardin du Roi à Paris." *Histoire et nature: Cahiers de l'Association pour l'histoire des sciences de la nature* 14 (1979): 3–29.

Chéruel, A. *Dictionnaire historique des institutions, moeurs et coutumes de la France.* 6th ed. Paris: Librairie Hachette, 1884.

——. *Histoire de l'administration monarchique en France depuis l'avènement de Philippe-Auguste jusqu'à la mort de Louis XIV.* 2 vols. Paris: Dezobry, E. Magdeleine et Cie, 1855.

Cipolla, Carlo M. "The Professions. The Long View." *The Journal of European Economic History* 2 (1973): 37–52.

Clamageran, Jean-Jules. *Histoire de l'impôt en France depuis l'époque romaine jusqu'à 1774.* 3 vols. Geneva: Slatkine Reprints, 1980 [1867].

Clarke, Jack A. "Abbé Jean-Paul Bignon 'Moderator of the Academies' and Royal Librarian." *French Historical Studies* 8 (1973): 213–35.

Collas, Georges. *Jean Chapelain, 1595–1674: Étude historique et littéraire d'après des documents inédits.* Paris: Perrin et Cie, 1912.

Costabel, Pierre. *Leibniz and Dynamics. The Texts of 1692.* Trans. R. E. W. Maddison. Ithaca: Cornell University Press; London: Methuen, 1973 [1960].

———. *Pierre Varignon (1654–1722) et la diffusion en France du calcul différentiel et intégral.* (Conférences du Palais de la découverte, D 108.) Paris: Université de Paris, 1965.

Crosland, Maurice, ed. *The Emergence of Science in Western Europe.* New York: Science History Publications, 1976.

Crousaz-Crétet, P. de. *Paris sous Louis XIV.* 2 vols. Paris: Librairie Plon, 1922–23.

Dictionnaire de biographie française. See Balteau et al.

Dictionary of Scientific Biography. See Gillispie, C. C.

Diderot, Denis. *L'Encyclopédie ou Dictionnaire raisonné des sciences, des arts et des métiers.* Facsimile of 1751–80 edition. Stuttgart-Bad Canstatt: Friedrich Fromann Verlag [Günther Holzboog], 1966.

Esnault, abbé G., and A. de Boislisle. "Documents relatifs à l'histoire des finances sous le règne de Louis XIV." *Bulletin du Comité des travaux historiques et scientifiques* (1883): 175–223.

Fabre, A. *Études littéraires sur le XVIIe siècle. Chapelain et nos deux premières Académies.* Paris: Perrin et Cie, 1890.

Forbonnais, François Véron de. *Recherches et considérations sur les finances de France, depuis l'année 1595 jusqu'à l'année 1721.* 2 vols. Basel: Cramer, 1758.

Franklin, Alfred. *Les anciennes bibliothèques de Paris. Églises, monastères, collèges, etc.* 3 vols. Histoire générale de Paris. Paris: Imprimerie impériale, 1867, 1870, 1873.

Gillispie, Charles Coulston, ed. *The Dictionary of Scientific Biography.* 16 vols. New York: Charles Scribner's Sons, 1970–80. (*DSB*).

Girault de Saint Fargeau, A. *Dictionnaire géographique, historique, industriel et commerciel de toutes les communes de la France.* 3 vols. Paris: Firmin Didot, 1844–46.

Goubert, Pierre. *Louis XIV and Twenty Million Frenchmen.* Trans. Anne Carter. New York: Pantheon Books, 1970.

Le grand vocabulaire français . . . , par une Société de gens de lettres. 30 vols. Paris: Panckoucke, 1767–74.

Grassby, R. B. "Social Status and Commercial Enterprise under Louis XIV." *The Economic History Review,* 2nd ser., 13 (1960): 19–38.

Greenberg, John L. "Mathematical Physics in Eighteenth-Century France." *Isis* 77 (1986): 59–78.

Guiffrey, Jules, ed. See *Les comptes des bâtiments.*

Hahn, Roger. *The Anatomy of a Scientific Institution: The Paris Academy of Sciences, 1666–1803.* Berkeley: University of California Press, 1971.

———. "Scientific Careers in Eighteenth-Century France," 127–38, in *The Emergence of Science in Western Europe,* ed. Maurice Crosland.

———. "Scientific Research as an Occupation in Eighteenth-Century Paris." *Minerva* 13 (1975): 501–13.

Handford, J. R. "Chemistry at the Jardin du roi from D'Avisson to Macquer." Unpublished M. Sc. thesis, University College, London, 1958.

Hauser, Henri. *Recherches et documents sur l'histoire des prix en France de 1500 à 1800.* Paris: Les presses modernes, 1936.

Hillairet, Jacques. *Dictionnaire historique des rues de Paris.* 3 vols. Paris: Les Éditions de minuit, n. d. [c. 1961].

Hoefer, Jean Chrétien Ferdinand. *Nouvelle biographie universelle.* 46 vols. Paris: Firmin Didot Frères, 1852–68. (*NBU*).

Howard, Rio. "Medical Politics and the Founding of the Jardin des Plantes in Paris." *Journal of the Society for the Bibliography of Natural History* 9 (1980): 395–402.

Huard, Georges. "Les planches de l'*Encyclopédie* et celles de *La description des arts et métiers* de l'Académie des Sciences," 35–46 in *L'Encyclopédie et le progrès des sciences et des techniques.* Paris: Presses Universitaires de France, 1952.

Imprimerie nationale. *L'art du livre à l'Imprimerie nationale.* Paris: Imprimerie nationale, 1973.

Institut de France. Académie des Sciences. *Index biographique des membres et correspondants de l'Académie des Sciences du 22 décembre 1666 au 15 décembre 1978.* Paris: Gauthier-Villars, 1979. (*IB*).

Jacq-Hergoualc'h, Michel. "Catalogue" in *Phra Narai, roi de Siam et Louis XIV.* See Ministère des Affaires Étrangères and Association Française d'Action Artistique.

Jacquiot, Josèphe. "Pourquoi l'Académie s'est-elle appelée Académie des Inscriptions." *Comptes rendus de l'Académie des Inscriptions et Belles-lettres* (1967): 134–49.

Jammes, André. "Académisme et Typographie: The Making of the Romain du Roi." *Journal of the Printing Historical Society* 1 (1965): 71–95. A slightly abridged translation of Jammes, *Réforme*, with the plates reduced.

———. "Innovation dans l'art de la lettre: Le Grandjean et la naissance de la typographie moderne," 127–41, in Imprimerie nationale, *L'art du livre.*

———. *La réforme de la typographie royale sous Louis XIV*. Paris: Librairie Paul Jammes, 1961. For a translation, see his "Académisme et Typographie."

Laissus, Yves and Anne-Marie Monseigny. "*Les Plantes du Roi:* Note sur un grand ouvrage de botanique préparé au XVIIe siècle par l'Académie royale des sciences." *Revue d'histoire des sciences* 22 (1969): 193–236.

Lavisse, Ernest. *Louis XIV.* 2 vols. Paris: Librairie Jules Tallandier, 1978 [1911].

Lefebvre, Georges. *The Coming of the French Revolution.* Trans. R. R. Palmer. Princeton: Princeton University Press, 1979 [1939].

Lehoux, Françoise. *Le cadre de vie des médecins parisiens aux XVIe et XVIIe siècles.* Paris: A. & J. Picard, 1976.

Lery, Edmond. "Le P. Sébastien Truchet en Auvergne." *Bulletin historique et scientifique de l'Auvergne* 57, no. 494 (1937): 49–58.

———. "Le P. Sébastien Truchet membre honoraire de l'Académie des sciences (1657–1729). Ses travaux à Versailles et à Marly." *Revue de l'histoire de Versailles et de Seine-et-Oise* (Oct.-Dec. 1929) (Versailles: Librairie Léon Bernard, 1929).

Malet [or Mallet], Jean Roland. *Comptes rendus de l'administration des finances du royaume de France, pendant les onze dernières années du règne de Henri IV, le règne de Louis XIII, & soixante-cinq années de celui de Louis XIV. . . .* Written in 1720. London and Paris: Buisson, 1789.

Martin, Henri-Jean. *Livre, pouvoirs et société à Paris au XVIIe siècle (1598–1701).* 2 vols. Histoire et civilisation du livre, Centre de recherches d'histoire et de philologie, 3. Geneva: Librairie Droz, 1969.

Maury, L.-F. Alfred. *Les Académies d'autrefois. L'ancienne Académie des inscriptions et belles-lettres.* Paris: Didier et Cie, 1864.

———. *Les Académies d'autrefois. L'ancienne Académie des sciences.* Paris: Didier et Cie, 1864.

McClellan, James E., III. "The Académie Royale des Sciences, 1699–1793: A Statistical Portrait." *Isis* 72 (1981): 541–67.

Mesnard, Jean. *Pascal et les Roannez.* 2 vols. n. p.: Desclée de Brouwer, 1965.

Meuvret, Jean. *Études d'histoire économique. Recueil d'articles.* Cahiers des Annales, 32. Paris: Librairie Armand Colin, 1971.

———. "Les mouvements des prix de 1661 à 1715 et leurs répercussions." *Journal de la Société de Statistique de Paris* 85 (1944): 109–19; reprinted in Meuvret, *Études.*

———. "La situation matérielle des membres du clergé séculier dans la France du XVIIe siècle." *Revue d'histoire de l'église de France* 1 (1968): 47–68; reprinted in Meuvret, *Études*, 251–68.

Middleton, W. E. Knowles. *The Experimenters: A Study of the Accademia del Cimento.* Baltimore: The Johns Hopkins Press, 1971.

Ministère de l'Industrie. "Carte des richesses minérales de la France" (1959).

Ministère de l'Industrie et du Commerce. "Carte géologique de la France." 4th ed. (1955).

Ministère des Affaires Étrangères and Association Française d'Action Artistique. *Phra Narai, roi de Siam et Louis XIV.* Paris: Musée de l'Orangerie, 13 June - 13 July 1986.

Mirot, Léon. *Roger de Piles. Peintre, amateur, critique, membre de l'Académie de peinture (1635–1709).* Paris: Jean Schemit, 1924.

Mols, Roger. *Introduction à la démographie historique des villes d'Europe du XVIe au XVIIe siècle.* 3 vols. Université de Louvain, Recueil de travaux d'histoire et de philologie, 4th ser., vols. 1–3. Louvain: Éditions J. Duculot, S. A. Gembloux, 1954–56.

Moréri, Louis. *Le grand dictionnaire historique.* 10 vols. Paris: Libraires Associés, 1759.

Mousnier, Roland. *Les institutions de la France sous la monarchie absolue, 1598–1789.* 2 vols. Paris: Presses Universitaires de France, 1974, 1980.

———. *Paris au XVIIe siècle.* Les Cours de Sorbonne. Histoire Certificat L. Paris: Centre de Documentation Universitaire [1961].

———. *Social Hierarchies 1450 to the Present.* Trans. Peter Evans, ed. Margaret Clarke. London: Croom Helm, 1973.

Neveu, Bruno. "La vie érudite à Paris à la fin du XVIIe siècle, d'après les papiers du P. Léonard de Sainte-Catherine (1695–1706)." *Bibliothèque de l'École des Chartes* 124 (1966): 432–511.

Nicéron, Jean Pierre. *Mémoires pour servir à l'histoire des hommes illustres dans la république des lettres.* 44 vols. Paris: Briasson, 1727–45.

Partington, J. R. *A History of Chemistry.* 4 vols. London: Macmillan; New York: St. Martin's Press, 1959–70.

Paul, Charles B. *Science and Immortality: The Éloges of the Paris Academy of Sciences (1699–1791).* Berkeley: University of California Press, 1980.

Peignot, Gabriel. *Documents authentiques et détails curieux sur les dépenses de Louis XIV.* Paris: Jules Renouard, Victor Lagier, 1827.

Pelletier, Monique. "Les globes de Louis XIV: Les sources françaises de l'oeuvre de Coronelli." *Imago Mundi* 34 (1982): 72–89.

Pingré, A.-G. *Annales célestes du dix-septième siècle.* Paris: Gauthier-Villars, 1901.

Pontal, Édouard. "La capitation en 1695," in Saint-Simon, *Mémoires,* 2: 458–68.

Proust, Jacques. *Diderot et l'Encyclopédie.* Paris: Armand Colin, 1962.

Rousselot de Surgy. "Discours préliminaire, ou Essai historique sur les finances," 1: i–lx in *Encyclopédie méthodique. Finances,* par une Société de gens de lettres. Paris: Panckoucke, 1784.

Robinet, André. "Le groupe malebranchiste introducteur du calcul infinitésimal en France." *Revue d'histoire des sciences* 13 (1960): 287–308.

Salomon-Bayet, Claire. "Un préambule théorique à une Académie des arts. Académie royale des sciences, 1693–1696. Présentation et textes." *Revue d'histoire des sciences* 23 (1970): 229–50.

Saunders, Elmo Stewart. *The Decline and Reform of the Académie des Sciences à Paris, 1676–1699.* Ann Arbor, Mich.: University Microfilms, 1981.

Schaeper, Thomas J. *The Economy of France in the Second Half of the Reign of Louis XIV.* Montreal: Interuniversity Centre for European Studies, 1980.

Sgard, Jean, Michel Gilot, and Françoise Weil, eds. *Dictionnaire des journalistes (1600–1789).* Grenoble: Presses Universitaires de Grenoble, 1976.

Steinberg, S. H. *Five Hundred Years of Printing.* Harmondsworth, Middlesex: Penguin Books, 1969 [1955, 1961].

Stroup, Alice. *A Company of Scientists: Botany, Patronage, and Community at the Seventeenth-Century Parisian Royal Academy of Sciences.* Berkeley: University of California Press, forthcoming.

——. "Wilhelm Homberg and the Search for the Constituents of Plants at the 17th-Century Académie Royale des Sciences." *Ambix* 26 (1979): 184–201.

Sturdy, David. "Pierre-Jean-Baptiste Chomel (1671–1740): A Case Study in Problems Relating to the Social Status of Scientists in the Early Modern Period." *The British Journal for the History of Science* 19 (1986): 301–22.

Symcox, Geoffry. *The Crisis of French Sea Power, 1688–1697. From the Guerre d'Escadre to the Guerre de Course.* Archives internationales d'histoire des idées, 73. The Hague: Martinus Nijhoff, 1974.

Teyssèdre, Bernard. *Roger de Piles et les débats sur le coloris au siècle de Louis XIV.* Paris: La Bibliothèque des Arts, 1957.

Ultee, Maarten. *The Abbey of St. Germain des Prés in the Seventeenth Century.* New Haven: Yale University Press, 1981.

Viard, Jules. "Les opérations du Bureau du triage. Notice et état sommaire de 11,760 liasses et registres de la Chambre des comptes détruits en l'an V." *Bibliothèque de l'École des Chartes* 57 (1896): 1–9.

Vührer, A. *Histoire de la dette publique en France.* Paris: Berger–Levrault et Cie, 1886.

Willems, H. and J.-Y. Conan. *Liste alphabétique des pages de la Grande écurie du Roi. Aperçu historique par M. le Vicomte Dominique Labarre de Raillicourt.* Dison-Verviens: Imprimerie Lelotte; Suresnes (Seine): M.-J.-Y. Cosson, n.d.

Wolf, Charles-Joseph-Etienne. *Histoire de l'Observatoire de Paris de sa fondation à 1793.* Paris: Gauthier-Villars, 1902.

Yates, Frances. *The French Academies of the Sixteenth Century.* Studies of the Warburg Institute, vol. 15. London: The Warburg Institute, University of London, 1947.

INDEX

Academicians, absenteeism, 30, 33, 34, 35–38, 84, 126–27; age at entry, 18n. 9; as part of the socioeconomic structure of France, 20–24; categories of membership, 14–18; foreign, 14, 15n. 6; needs and responsibilities of, 10; other income, 25–27; students, 40; unpaid, 15–18, 81, 40. *See also* Pensions; names of individual academicians

Académie française, 6n. 22, 20n. 3, 103n. f

Académie royale des inscriptions, 7, 21, 23, 28, 29n. 39, 30nn. 41, 44, 41, 70n. 12, 75–76, 79; pensions of members, 69–70, 71, 79–81, 85, 90, 91, 93, 94, 96, 97, 99, 100, 100n. d, 102, 103, 105, 106, 108, 109, 111, 112, 115, 117, 118, 121. *See also Petite compagnie; Règlement* of 1701; names of individual academicians

Académie royale des sciences. *See also* Academicians; Académie royale des inscriptions; Compagnie des arts et métiers; Expeditions; Jesuit missionary-scientists; Laboratory; Mathematical sciences; Mathematics; Ministerial supervision; Natural history; Natural philosophy; Patronage; Pensions; Utilitarian interests
—assistants to, 18n. 11, 46n. 28, 55; their pensions, 80, 81, 90, 91, 91n. b, 93, 94, 96, 97, 99, 100, 102, 103, 105, 106, 108, 109, 111, 112, 114, 115, 117, 118, 120, 121, 122, 152. *See also* Chastillon; Dalesme; "Garçon cirugien"
—development of in 1690s, 9, 39
—finances, 1, 2, 4–6, 7, 8, 10, 11, 40–50, 59, 61–63; Appendices A, B, C; Tables 1–8
—functions of, 4–5, 53–60, 61, 63
—history and traditions of, 14n. 2, 37, 38, 40, 58n. 28
—meetings of, 4n. 13, 33–38
—morale of, 10, 11, 30–31, 33–34, 35–38, 50
—planning at, 33, 34
—publications of, 4, 5, 6, 10, 11, 39, 40, 44–45, 47, 48, 49, 50–51, 53, 54; importance of in Pontchartrain's program, 40, 50–51, 53, 60. *See also* Illustrations
—reorganization of in 1699. *See Règlement* of 1699
—research, 33, 39–40, 51; cost of, 11, 40–50, 61. *See also* Expeditions; Mathematical sciences; Mathematics; Natural philosophy; Utilitarian interests
—size of, 11–14, 40; working Academy, 33, 35, 38

Academies, royal, 5, 33n. *See also* Académie française; Académie royale des inscriptions; Académie royale des sciences

Academy of Inscriptions. *See* Académie royale des inscriptions

Academy of Sciences. *See* Académie royale des sciences

Affaires secrettes, 72

Air, elasticity and pressure of, 39

Ambassadors, 72

Anatomists, 50; attendance of, 36. *See also* names of individual academicians

Anatomy, 34, 37, 40, 41, 46, 47–48, 49–50, 63, 88, 132–33; summary of cost, 149, 150–51. *See also* Anatomists; Natural history; Natural philosophy; names of individual academicians

Animals. *See* Anatomy; Illustrations; Natural history

Anisson, 58, 58nn. 29, 31

Anjou, duke of, 27n. 27, 55

Annuities, 72, 73; advantages and disadvantages of, 29–30. *See also Rentes*

Apparatus. *See* Instruments, scientific

Appendix A, 65–72; 7nn. 24, 26, 8, 28n. 34, 118n. a

Appendix B, 73–78; 8, 27n. 27, 28, 28nn. 32, 34, 29n. 39, 30nn. 41, 42, 45n. 23, 54n. 6, 70, 71, n. 15, 72, 103n. c, 125n. 4

Appendix C, 79–85; 7n. 25, 8, 13n. c, 17n. c, 19n. 12, 59n. 32, 100n. b, 118n. c, 122n. a

Archives Nationales, 7, 20n. 2, 65

Artisans, interviews with, 60; paid for their information, 59

Assignations, 71

Associate academicians, 45

Astronomers, 38, 43, 44, 45, 46, 77–78; attendance of, 37. *See also* names of individual academicians

Astronomy, 39, 40, 43, 44, 45, 47, 53. *See also* Observatory; names of individual academicians

Augmentations, 27, 75, 77, 119, 121n, 125nn. 3, 4

Auvergne, 49

Auzout, Adrien, 12, 13n. b, 16, 48n. 33

Baleuze, 124

Barabel, paid, 146–47

Beaupré, 125

Benoit, Jean, paid, 132–33

160

PUBLICATIONS

OF

The American Philosophical Society

The publications of the American Philosophical Society consist of PROCEEDINGS, TRANSACTIONS, MEMOIRS, and YEAR BOOK.

THE PROCEEDINGS contains papers which have been read before the Society in addition to other papers which have been accepted for publication by the Committee on Publications. In accordance with the present policy one volume is issued each year, consisting of four numbers, and the price is $20.00 net per volume. Individual copies of the PROCEEDINGS are $10.00.

THE TRANSACTIONS, the oldest scholarly journal in America, was started in 1769. In accordance with the present policy each annual volume is a collection of monographs, each issued as a part. The current annual subscription price is $60.00 net per volume. Individual copies of the TRANSACTIONS are offered for sale.

Each volume of the MEMOIRS is published as a book. The titles cover the various fields of learning; most of the recent volumes have been historical. The price of each volume is determined by its size and character, but subscribers are offered a 20 per cent discount.

The YEAR BOOK is of considerable interest to scholars because of the reports on grants for research and to libraries for this reason and because of the section dealing with the acquisitions of the Library. In addition it contains the Charter and Laws, and lists of members, and reports of committees and meetings. The YEAR BOOK is published about April 1 for the preceding calendar year. The current price is $8.50.

An author desiring to submit a manuscript for publication should send it to the Editor, American Philosophical Society, 104 South Fifth Street, Philadelphia, Pa. 19106.

www.ingramcontent.com/pod-product-compliance
Lightning Source LLC
Chambersburg PA
CBHW080932290326
41932CB00041B/2